# CHEMISTRY: INORGANIC,
# ORGANIC AND BIOLOGICAL

## About the Author

Philip S. Chen received his Ph.D. degree in organic chemistry from Michigan State University in 1933. For five years he taught at Madison College. Since 1938 he has been associated with Atlantic Union College, where he is now Professor of Chemistry, Head of the Chemistry Department, and Chairman of the Division of Biology and Chemistry.

Professor Chen has been a contributor to the *Journal of the American Chemical Society*, *Journal of Chemical Education*, *Chemist Analyst*, *Journal of American Leather Chemists Association*, and the *Chinese Medical Journal*. He is also the author of the widely used *Chemical Elements Wall Chart*, as well as five books related to science. He is a member of the American Chemical Society, the American Association for the Advancement of Science, the New York Academy of Sciences, and Sigma Xi.

# CHEMISTRY:
## INORGANIC
## ORGANIC and
## BIOLOGICAL

Philip S. Chen

**IIIII BARNES & NOBLE BOOKS**

A DIVISION OF HARPER & ROW, PUBLISHERS

New York, Hagerstown, San Francisco, London

Printed in the United States of America

79 80   12 11

# PREFACE

This book was written to provide a study-guide for students taking a course covering all three important fields of chemistry—inorganic, organic, and biological. Some of these students will be preparing for a nursing career; others will be students of home economics, physical education, agriculture, or biology; still others will be taking a chemistry course to fulfill their science requirement or to obtain a general background in chemistry. To help these students absorb the basic concepts and facts from these three fields of chemistry, this book summarizes the fundamental material in as clear and readable a form as possible.

The author is indebted to his colleagues Professors Charles W. Slattery and Leonard L. Nelson for their helpful criticism and constructive suggestions. He also wishes to thank Mrs. Jeanne Flagg of the editorial staff of Barnes & Noble for her efficient and expert editing.

Philip S. Chen

South Lancaster, Mass.

# CONTENTS

# CONTENTS

# PERIODIC TABLE OF THE ELEMENTS

| IA | IIA | IIIB | IVB | VB | VIB | VIIB | VIII | | | IB | IIB | IIIA | IVA | VA | VIA | VIIA | O |
|---|---|---|---|---|---|---|---|---|---|---|---|---|---|---|---|---|---|
| 1 **H** 1.00797 | | | | | | | | | | | | | | | | | 2 **He** 4.0026 |
| 3 **Li** 6.939 | 4 **Be** 9.0122 | | | | | | | | | | | 5 **B** 10.811 | 6 **C** 12.01115 | 7 **N** 14.0067 | 8 **O** 15.9994 | 9 **F** 18.9984 | 10 **Ne** 20.183 |
| 11 **Na** 22.9898 | 12 **Mg** 24.312 | | | | | | | | | | | 13 **Al** 26.9815 | 14 **Si** 28.086 | 15 **P** 30.9738 | 16 **S** 32.064 | 17 **Cl** 35.453 | 18 **Ar** 39.948 |
| 19 **K** 39.102 | 20 **Ca** 40.08 | 21 **Sc** 44.956 | 22 **Ti** 47.90 | 23 **V** 50.942 | 24 **Cr** 51.996 | 25 **Mn** 54.9380 | 26 **Fe** 55.847 | 27 **Co** 58.9332 | 28 **Ni** 58.71 | 29 **Cu** 63.54 | 30 **Zn** 65.37 | 31 **Ga** 69.72 | 32 **Ge** 72.59 | 33 **As** 74.9216 | 34 **Se** 78.96 | 35 **Br** 79.909 | 36 **Kr** 83.80 |
| 37 **Rb** 85.47 | 38 **Sr** 87.62 | 39 **Y** 88.905 | 40 **Zr** 91.22 | 41 **Nb** 92.906 | 42 **Mo** 95.94 | 43 **Tc** (99) | 44 **Ru** 101.07 | 45 **Rh** 102.905 | 46 **Pd** 106.4 | 47 **Ag** 107.870 | 48 **Cd** 112.40 | 49 **In** 114.82 | 50 **Sn** 118.69 | 51 **Sb** 121.75 | 52 **Te** 127.60 | 53 **I** 126.9044 | 54 **Xe** 131.30 |
| 55 **Cs** 132.905 | 56 **Ba** 137.34 | 57 ***La** 138.91 | 72 **Hf** 178.49 | 73 **Ta** 180.948 | 74 **W** 183.85 | 75 **Re** 186.2 | 76 **Os** 190.2 | 77 **Ir** 192.2 | 78 **Pt** 195.09 | 79 **Au** 196.967 | 80 **Hg** 200.59 | 81 **Tl** 204.37 | 82 **Pb** 207.19 | 83 **Bi** 208.980 | 84 **Po** (210) | 85 **At** (210) | 86 **Rn** (222) |
| 87 **Fr** (223) | 88 **Ra** (226) | 89 †**Ac** (227) | | | | | | | | | | | | | | | |

*Lanthanum Series

| 58 **Ce** 140.12 | 59 **Pr** 140.907 | 60 **Nd** 144.24 | 61 **Pm** (145) | 62 **Sm** 150.35 | 63 **Eu** 151.96 | 64 **Gd** 157.25 | 65 **Tb** 158.924 | 66 **Dy** 162.50 | 67 **Ho** 164.930 | 68 **Er** 167.26 | 69 **Tm** 168.934 | 70 **Yb** 173.04 | 71 **Lu** 174.97 |
|---|---|---|---|---|---|---|---|---|---|---|---|---|---|

†Actinium Series

| 90 **Th** 232.038 | 91 **Pa** (231) | 92 **U** 238.03 | 93 **Np** (237) | 94 **Pu** (242) | 95 **Am** (243) | 96 **Cm** (247) | 97 **Bk** (249) | 98 **Cf** (251) | 99 **Es** (254) | 100 **Fm** (253) | 101 **Md** (256) | 102 **No** (253) | 103 **Lw** (257) |
|---|---|---|---|---|---|---|---|---|---|---|---|---|---|

The numbers in parentheses are the mass numbers of most stable or most common isotope.

# ATOMIC WEIGHTS

| | Symbol | Atomic Number | Atomic Weight* | | Symbol | Atomic Number | Atomic Weight* |
|---|---|---|---|---|---|---|---|
| Actinium | Ac | 89 | [227] | Mercury | Hg | 80 | 200.59 |
| Aluminum | Al | 13 | 26.9815 | Molybdenum | Mo | 42 | 95.94 |
| Americium | Am | 95 | [243] | Neodymium | Nd | 60 | 144.24 |
| Antimony | Sb | 51 | 121.75 | Neon | Ne | 10 | 20.183 |
| Argon | Ar | 18 | 39.948 | Neptunium | Np | 93 | [237] |
| Arsenic | As | 33 | 74.9216 | Nickel | Ni | 28 | 58.71 |
| Astatine | At | 85 | [210] | Niobium | Nb | 41 | 92.906 |
| Barium | Ba | 56 | 137.34 | Nitrogen | N | 7 | 14.0067 |
| Berkelium | Bk | 97 | [247] | Nobelium | No | 102 | [254] |
| Beryllium | Be | 4 | 9.0122 | Osmium | Os | 76 | 190.2 |
| Bismuth | Bi | 83 | 208.980 | Oxygen | O | 8 | 15.9994 |
| Boron | B | 5 | 10.811 | Palladium | Pd | 46 | 106.4 |
| Bromine | Br | 35 | 79.904 | Phosphorus | P | 15 | 30.9738 |
| Cadmium | Cd | 48 | 112.40 | Platinum | Pt | 78 | 195.09 |
| Calcium | Ca | 20 | 40.08 | Plutonium | Pu | 94 | [244] |
| Californium | Cf | 98 | [251] | Polonium | Po | 84 | [209] |
| Carbon | C | 6 | 12.01115 | Potassium | K | 19 | 39.102 |
| Cerium | Ce | 58 | 140.12 | Praeseodymium | Pr | 59 | 140.907 |
| Cesium | Cs | 55 | 132.905 | Promethium | Pm | 61 | [145] |
| Chlorine | Cl | 17 | 35.453 | Protactinium | Pa | 91 | [231] |
| Chromium | Cr | 24 | 51.996 | Radium | Ra | 88 | [226] |
| Cobalt | Co | 27 | 58.9332 | Radon | Rn | 86 | [222] |
| Copper | Cu | 29 | 63.546 | Rhenium | Re | 75 | 186.2 |
| Curium | Cm | 96 | [247] | Rhodium | Rh | 45 | 102.905 |
| Dysprosium | Dy | 66 | 162.50 | Rubidium | Rb | 37 | 85.47 |
| Einsteinium | Es | 99 | [254] | Ruthenium | Ru | 44 | 101.07 |
| Erbium | Er | 68 | 167.26 | Samarium | Sm | 62 | 150.35 |
| Europium | Eu | 63 | 151.96 | Scandium | Sc | 21 | 44.956 |
| Fermium | Fm | 100 | [257] | Selenium | Se | 34 | 78.96 |
| Fluorine | F | 9 | 18.9984 | Silicon | Si | 14 | 28.086 |
| Francium | Fr | 87 | [223] | Silver | Ag | 47 | 107.868 |
| Gadolinium | Gd | 64 | 157.25 | Sodium | Na | 11 | 22.9898 |
| Gallium | Ga | 31 | 69.72 | Strontium | Sr | 38 | 87.62 |
| Germanium | Ge | 32 | 72.59 | Sulfur | S | 16 | 32.064 |
| Gold | Au | 79 | 196.967 | Tantalum | Ta | 73 | 180.948 |
| Hafnium | Hf | 72 | 178.49 | Technetium | Tc | 43 | [97] |
| Helium | He | 2 | 4.0026 | Tellurium | Te | 52 | 127.60 |
| Holmium | Ho | 67 | 164.930 | Terbium | Tb | 65 | 158.924 |
| Hydrogen | H | 1 | 1.00797 | Thallium | Tl | 81 | 204.37 |
| Indium | In | 49 | 114.82 | Thorium | Th | 90 | 232.038 |
| Iodine | I | 53 | 126.9004 | Thulium | Tm | 69 | 168.934 |
| Iridium | Ir | 77 | 192.2 | Tin | Sn | 50 | 118.69 |
| Iron | Fe | 26 | 55.847 | Titanium | Ti | 22 | 47.90 |
| Krypton | Kr | 36 | 83.80 | Tungsten | W | 74 | 183.85 |
| Lanthanum | La | 57 | 138.91 | Uranium | U | 92 | 238.03 |
| Lawrencium | Lw | 103 | [257] | Vanadium | V | 23 | 50.942 |
| Lead | Pb | 82 | 207.19 | Xenon | Xe | 54 | 131.30 |
| Lithium | Li | 3 | 6.939 | Ytterbium | Yb | 70 | 173.04 |
| Lutetium | Lu | 71 | 174.97 | Yttrium | Y | 39 | 88.905 |
| Magnesium | Mg | 12 | 24.312 | Zinc | Zn | 30 | 65.37 |
| Manganese | Mn | 25 | 54.9380 | Zirconium | Zr | 40 | 91.22 |
| Mendelevium | Md | 101 | [256] | | | | |

*A value given in brackets denotes the mass number of the isotope of longest known half-life.

# CHEMISTRY: INORGANIC, ORGANIC AND BIOLOGICAL

# 1. SOME FUNDAMENTAL CONCEPTS

Chemistry is the science that deals with the composition and properties of substances and the transformations they undergo.

## DIVISIONS OF CHEMISTRY

*General chemistry* is a broad survey of chemistry as a whole, with special emphasis on its basic principles and laws. It includes the properties and reactions of some of the most common elements and compounds.

*Organic chemistry* is the study of compounds of carbon, either as they are produced in plants and animals or as they are formed synthetically.

*Biochemistry* is the study of the chemistry of living processes. All the chemical reactions taking place in the body are more specifically referred to as *physiological chemistry*.

*Analytical chemistry* is concerned with the methods of determining the various constituents of matter as to *what* they are (*qualitative analysis*), or *how much* they are (*quantitative analysis*).

*Physical chemistry* deals with the principles and laws that underlie chemical changes.

*Nuclear chemistry* is the study of changes that take place in the nucleus of the atom.

## MATTER

Matter is anything that occupies space and has mass. In ordinary chemical reactions, matter can neither be created nor destroyed (Law of Conservation of Matter). In nuclear reactions, however, matter can be converted into energy, or vice versa. These conversions will be discussed in Chapter 4.

**Physical States of Matter.** Matter exists in three physical states: solid, liquid, and gaseous, depending on temperature and pressure. *Solids* are rigid and have a definite volume and a definite form. *Liquids* have a definite volume but no definite form. They flow and assume the shape of the vessel which holds

1

them. *Gases* have neither a definite volume nor a definite form. They diffuse into every part of the container in which they are placed.

**Classification of Matter.** Matter can be in the form of an element, a compound, or a mixture.

ELEMENTS. An element is a substance that cannot be decomposed into simpler substances by ordinary chemical means. It may also be defined as a substance whose properties give it a definite place in the periodic table. There are 103 known chemical elements at the present time. They may be classified into metals and nonmetals. Examples of metals are iron, silver, and gold. Sulfur, oxygen, and nitrogen are nonmetals.

COMPOUNDS. A compound is made up of two or more elements chemically combined in definite proportions by weight. Thus, the compound water is composed of 11.11 percent hydrogen and 88.89 percent oxygen by weight. A compound is homogeneous. Its properties are quite different from those of its constituent elements, and its constituent elements can be separated only by chemical means. The definite ratio by weight of hydrogen to oxygen in water illustrates the Law of Definite Proportions, or the Law of Definite Composition.

MIXTURES. A mixture is made up of two or more substances that are not combined chemically. Its component parts retain their own properties and can be separated by mechanical means. For example, cream of tartar baking powder is a mixture of sodium bicarbonate, cream of tartar, and starch.

**Changes in Matter.** Matter undergoes changes, some of which are physical and others of which are chemical.

PHYSICAL CHANGES. A physical change is an alteration in the condition or state of a substance. The chemical composition of the substance is not changed. Examples of physical changes are the chopping of wood, the breaking of glass, and the melting of ice.

CHEMICAL CHANGES. A chemical change is one in which a new substance is formed having a composition and properties different from those of the original substance. For example, iron on exposure to moist air becomes rust; sulfur, on burning, becomes sulfur dioxide.

**Properties of Matter.** The distinguishing characteristics of a substance are referred to as its properties. There are two types of properties: physical and chemical.

PHYSICAL PROPERTIES. The physical properties of a substance are those associated with physical changes. They include characteristics such as color, odor, taste, density, crystalline form, boiling point, and melting point.

CHEMICAL PROPERTIES. Chemical properties are characteristics of elements and compounds which describe the manner in which these substances react with other substances. For example, sodium reacts readily with water to liberate hydrogen; water is decomposed into hydrogen and oxygen by electrolysis.

## ENERGY

Energy may be defined as the ability to do work. Matter always possesses energy in one form or another. All transformations of matter are accompanied by transformations of energy.

**Forms of Energy and Transformation.** Energy can take many forms: heat, light, electrical, kinetic, chemical, and nuclear or atomic energy. Energy can be changed from one form to another but cannot be created or destroyed in reactions other than nuclear reactions (Law of Conservation of Energy). For example, electricity is changed into light in a light bulb; heat from steam is changed into electricity in an electric generator.

Every chemical change is accompanied by a change in energy. When carbon is oxidized to carbon dioxide, chemical energy is converted into heat energy. When magnesium burns in air to magnesium oxide, chemical energy is changed into light and heat.

**Heat Energy.** The unit for measuring heat energy is the *calorie*. The small calorie (cal.) is the amount of heat required to raise the temperature of one gram of water one degree Centigrade. The large Calorie or kilocalorie (Cal. or Kg.-Cal.) is the amount of heat required to raise the temperature of one kilogram (1,000 grams) of water one degree Centigrade. *Specific heat* is the number of calories required to raise the temperature of one gram of a substance one degree Centigrade.

**The Kinetic Molecular Theory.** The energy of motion is called *kinetic energy*. The kinetic molecular theory was proposed to explain the three states of matter. The following assumptions were made to account for the behavior of gases.

1. All gases are composed of molecules, which are relatively far apart. This explains the compressibility of gases.

2. The molecules of gases move rapidly in all directions, which explains the diffusion of gases.

3. Gas molecules collide with each other and with the walls of the container. The pressure which gases exert is the result of the individual fast-moving molecules striking the sides of the container.

4. The velocities of gas molecules increase when the temperature is increased and decrease when the temperature is decreased. This explains, for example, why the pressure inside automobile tires is increased when traveling over hot roads.

From this theory it was deduced that as a gas is cooled, its kinetic energy decreases and it becomes a liquid. If further cooled, a solid is formed. At absolute zero ($-273°C.$), theoretically there is no molecular motion at all.

## THE METRIC SYSTEM

Most countries of the world use the metric system, especially in scientific work. The advantage of the metric system is that the various units referring to a given dimension differ from one another by multiples of ten. The standard of length is the meter, of weight the kilogram, and of volume the liter. The most frequently used units are given below.

### UNITS OF LENGTH

| | | | |
|---|---|---|---|
| Micron ($\mu$) | = | 0.001 | millimeter |
| Millimeter (mm.) | = | 0.001 | meter |
| Centimeter (cm.) | = | 0.01 | meter |
| Meter (m.) | = | 1.0 | meter |
| Kilometer (km.) | = | 1000.0 | meters |

### UNITS OF WEIGHT

| | | | |
|---|---|---|---|
| Microgram ($\mu$g.) | = | 0.001 | milligram |
| Milligram (mg.) | = | 0.001 | gram |
| Gram (gm.) | = | 1.0 | gram |
| Kilogram (kg.) | = | 1000.0 | grams |

### UNITS OF VOLUME

| | | | |
|---|---|---|---|
| Milliliter (ml.) | = | 0.001 | liter |
| Liter (l.) | = | 1.0 | liter |

### SOME APPROXIMATE EQUIVALENTS

| | | | |
|---|---|---|---|
| 1 | inch | = | 2.5 | cm. |
| 1 | foot | = | 30 | cm. |
| 1 | ounce | = | 28 | gm. |
| 1 | pound | = | 454 | gm. |
| 2.2 | pounds | = | 1 | kg. |
| 1 | cubic centimeter | = | 1 | ml. |
| 1 | fluidounce | = | 30 | ml. |

| 1 | quart | = | 946 | ml. |
|---|-------|---|-----|-----|
| 1 | cup | = | 250 | ml. |
| 1 | teaspoon | = | 4 | ml. |
| 1 | tablespoon | = | 15 | ml. |

## MEASUREMENT OF TEMPERATURE

Standards of temperature are based on the freezing point and the boiling point of water. On the Centigrade scale, the freezing point of water is set as 0° and the boiling point as 100°. The Fahrenheit scale has the freezing point as 32° and the boiling point as 212°. The Absolute (Kelvin) scale reads 273° higher than the Centigrade. The three scales are compared in Fig. 1.1.

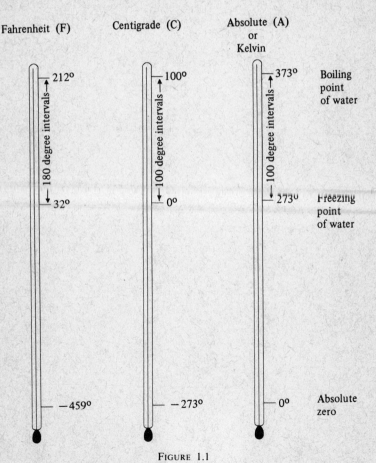

FIGURE 1.1

The formulas used for converting temperatures from one scale to another are as follows:

**To convert Fahrenheit to centigrade**
Method 1: $C = \frac{5}{9}(F - 32)$
Method 2: $C = \frac{5}{9}(F + 40) - 40$

**To convert centigrade to Fahrenheit**
Method 1: $F = \frac{9}{5}C + 32$
Method 2: $F = \frac{9}{5}(C + 40) - 40$

**To convert centigrade to absolute**
$A = C + 273$

# 2. ATOMIC STRUCTURE

According to John Dalton's atomic theory, molecules were made up of atoms which were hard, indivisible particles. Later studies by other scientists, using X rays, radioactivity, and spectroscopic analysis, revealed, however, that atoms are not indivisible, but consist of positively charged nuclei with electrons revolving around them.

## FUNDAMENTAL UNITS OF THE ATOM

The atom is made up of three kinds of particles: protons, electrons, and neutrons.

**The Proton.** A proton is a particle with a positive charge of 1. It has about the same mass as that of the hydrogen atom. Protons are located in the nucleus of the atom.

**The Electron.** An electron is a particle with a negative charge of 1. Its mass is about 1/1837 that of the hydrogen atom. Electrons are located in the electron shells, or energy levels, outside the nucleus.

**The Neutron.** A neutron has about the same mass as that of a proton but has no electrical charge. Like the proton, it is located in the nucleus of the atom.

*Examples:* The hydrogen atom has one proton (+) in the nucleus surrounded by one electron (· or e⁻) in the shell outside the nucleus. The helium atom has two protons and two neutrons (±) in the nucleus surrounded by two electrons in the outside shell.

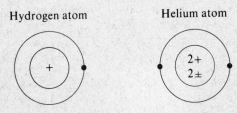

Hydrogen atom          Helium atom

# ATOMIC STRUCTURE AND CHEMICAL PROPERTIES

The chemical properties of elements are closely related to their atomic structure. The study of atomic structure will therefore enable us to explain, for example, why certain elements are inert while others are reactive; why certain elements are metals while others are nonmetals; why one atom of one element can react with one, two, or three atoms of another element.

The *atomic number* is the number of protons in the nucleus of an atom. The hydrogen atom has one proton; therefore its atomic number is one. The oxygen atom has eight protons, so its atomic number is eight.

In the neutral atom, the total number of electrons distributed in the various shells, or energy levels, outside the nucleus is equal to the number of protons within the nucleus.

There are seven main energy levels in which electrons move. The maximum number of electrons that each shell can have is as follows:

| Electron shell or energy level* | K | L | M | N | O | P | Q |
|---|---|---|---|---|---|---|---|
| Maximum number of electrons | 2 | 8 | 18 | 32 | 32 | 18 | 8 |

M, N, O, and P shells cannot have the maximum number of electrons indicated above if there are no other shells outside them. In the outermost shell the maximum number of electrons that can be present is eight. See Table 2.1.

Elements which have two electrons in the K shell or eight electrons in any of the other outermost shells are chemically inert and will not react with other substances under ordinary conditions. The Group O elements† are of this type; these are the *noble*, or inert, gases helium, neon, argon, xenon, etc. However, by using modern techniques, a number of noble gas compounds have been prepared, especially those of xenon.

Group I to Group III elements have one to three electrons, respectively, in the outermost shell. They can lose these electrons in order to assume the more stable atomic configuration of the noble gases, that is, two or eight electrons (octet) in the outermost

---

*The energy levels, represented here by letters, may also be designated by the numbers 1, 2, 3, 4, 5, 6, and 7.

†The elements can be classified into groups according to the periodic law, which will be discussed in Chapter 3.

## TABLE 2.1. ELECTRON CONFIGURATIONS OF SOME REPRESENTATIVE ELEMENTS

| Atomic No. | Element | Electron Configuration | | | | | | |
|---|---|---|---|---|---|---|---|---|
| | | K | L | M | N | O | P | Q |
| 1 | Hydrogen | 1 | | | | | | |
| 2 | Helium | 2 | | | | | | |
| 3 | Lithium | 2 | 1 | | | | | |
| 4 | Beryllium | 2 | 2 | | | | | |
| 5 | Boron | 2 | 3 | | | | | |
| 6 | Carbon | 2 | 4 | | | | | |
| 7 | Nitrogen | 2 | 5 | | | | | |
| 8 | Oxygen | 2 | 6 | | | | | |
| 9 | Fluorine | 2 | 7 | | | | | |
| 10 | Neon | 2 | 8 | | | | | |
| 11 | Sodium | 2 | 8 | 1 | | | | |
| 12 | Magnesium | 2 | 8 | 2 | | | | |
| 13 | Aluminum | 2 | 8 | 3 | | | | |
| 14 | Silicon | 2 | 8 | 4 | | | | |
| 15 | Phosphorus | 2 | 8 | 5 | | | | |
| 16 | Sulfur | 2 | 8 | 6 | | | | |
| 17 | Chlorine | 2 | 8 | 7 | | | | |
| 18 | Argon | 2 | 8 | 8 | | | | |
| 19 | Potassium | 2 | 8 | 8 | 1 | | | |
| 29 | Copper | 2 | 8 | 18 | 1 | | | |
| 47 | Silver | 2 | 8 | 18 | 18 | 1 | | |
| 79 | Gold | 2 | 8 | 18 | 32 | 18 | 1 | |
| 103 | Lawrencium | 2 | 8 | 18 | 32 | 32 | 9 | 2 |

shell. The atoms that have lost electrons become positively charged ions.

*Examples:*

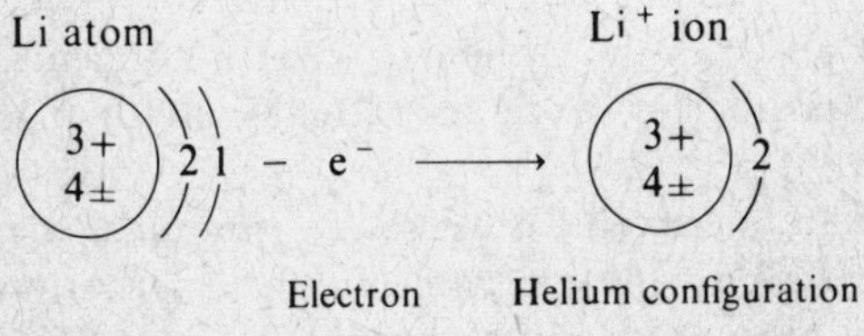

Li atom → Li$^+$ ion

$$3+ \quad 4\pm \quad 2\;1 \; - \; e^- \longrightarrow \; 3+ \quad 4\pm \quad 2$$

Electron   Helium configuration

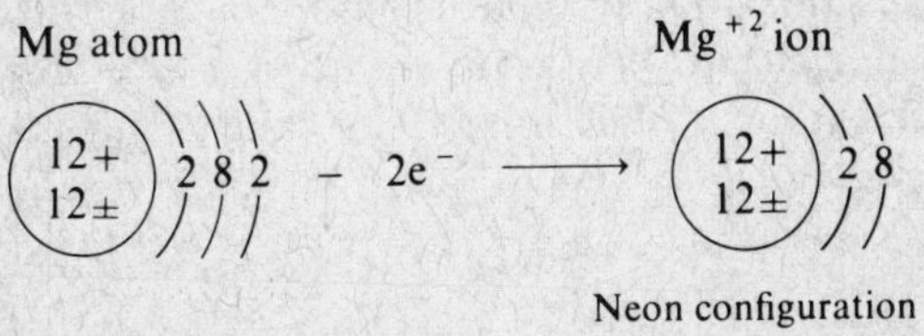

Mg atom → Mg$^{+2}$ ion

$$12+ \quad 12\pm \quad 2\;8\;2 \; - \; 2e^- \longrightarrow \; 12+ \quad 12\pm \quad 2\;8$$

Neon configuration

Elements in Group VIA have six electrons in their outermost shell and can gain two more electrons. In order to complete the octet and assume the atomic configuration of the noble gases,

Group VIIA elements, with seven electrons, can gain one more electron. The atoms that have gained electrons become negatively charged ions.

O atom                                              $O^{-2}$ ion

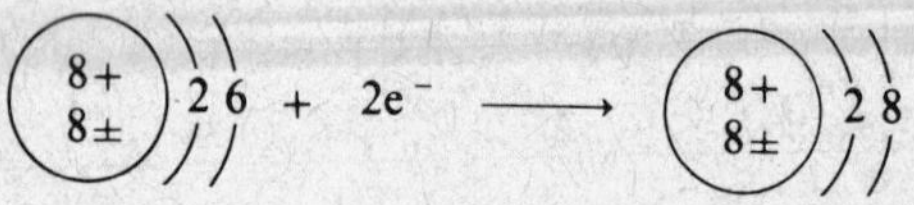

$$\left(\begin{array}{c} 8+ \\ 8\pm \end{array}\right) 2\ 6 \ + \ 2e^- \longrightarrow \left(\begin{array}{c} 8+ \\ 8\pm \end{array}\right) 2\ 8$$

Neon configuration

F atom                                              $F^-$ ion

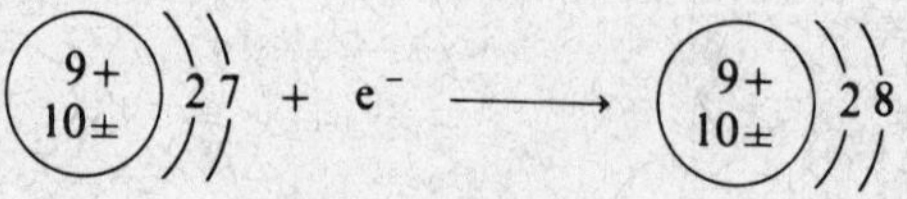

$$\left(\begin{array}{c} 9+ \\ 10\pm \end{array}\right) 2\ 7 \ + \ e^- \longrightarrow \left(\begin{array}{c} 9+ \\ 10\pm \end{array}\right) 2\ 8$$

Neon configuration

## ATOMIC WEIGHT

The atomic weight of an element may be considered to be the sum of the protons and neutrons in its nucleus. Thus, helium having 2 protons and 2 neutrons in the nucleus has the atomic weight of 4, and oxygen having 8 protons and 8 neutrons in the nucleus has the atomic weight of 16. A table of atomic weights will be found following the table of contents.

The atomic weight of an element may also be defined as its relative weight as compared to the weight of carbon which was arbitrarily fixed at 12 as the standard. The hydrogen atom, which weighs 1/12 as much as the carbon atom, therefore has the atomic weight of 1.

When the atomic weight is expressed in grams, it is called the *gram-atomic weight*. In one gram-atomic weight of any element, there are the same number of atoms as in one gram-atomic weight of any other element. A gram atom of any element contains $6.023 \times 10^{23}$ atoms. This number is called *Avogadro's number*. For example, in 12 grams of carbon, there are $6.023 \times 10^{23}$ atoms.

## ISOTOPES

Isotopes are different forms of the same element having an equal number of protons but an unequal number of neutrons.

The difference in the number of neutrons gives the isotopes a different mass; therefore they have different atomic weights. The two common isotopes of chlorine are shown below.

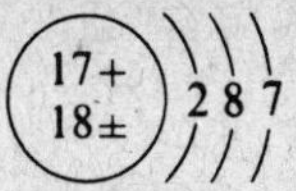

Chlorine 35 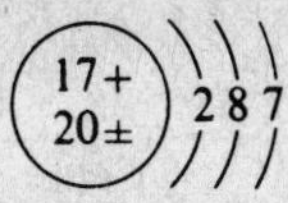      Chlorine 37

Ordinary chlorine contains both kinds of atoms. Consequently, the atomic weight 35.453 represents the average weight of the two different kinds of chlorine atoms as they occur in natural chlorine.

Isotopes may be considered, therefore, as atoms of the same element having the same atomic number but different atomic weights. They have the same chemical properties.

# 3. THE PERIODIC TABLE AND THE PERIODIC LAW

The periodic table is an attempt to classify the 103 known chemical elements according to their atomic weights and atomic numbers. By arranging all the elements in the order of their increasing atomic numbers, we find that the properties of the elements recur at regular intervals. For example, the melting points of the elements increase from hydrogen to carbon, then decrease, then increase again until silicon, then decrease again, then increase again until chromium, etc. See Fig. 3.1.

The melting points and other properties of the elements obey the *periodic law*, which may be stated: The properties of the elements are periodic functions of their atomic numbers.

There are two main forms of the periodic table, that of Mendeleev and the so-called long form.

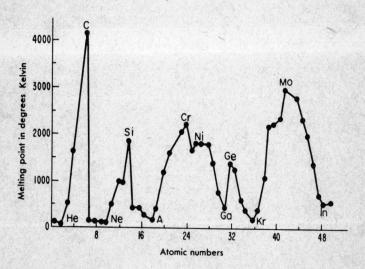

FIG. 3.1. The periodic law illustrated by a plot of atomic numbers vs. melting points of the elements, from atomic numbers 1 to 50. (From *General Chemistry*, 4th ed., by P. W. Selwood. Copyright 1950, 1954, © 1959, 1965 by Holt, Rinehart and Winston, Inc.)

## MENDELEEV'S PERIODIC TABLE

This table was devised in 1869 by the Russian chemist Dmitri Mendeleev, who arranged the elements in the order of their increasing atomic weights. Because the elements in subgroups A and B, such as potassium of group IA and copper of group IB, occupy the same vertical column instead of separate columns, it is not as useful as the long form.

## THE LONG FORM PERIODIC TABLE

This form of the periodic table is based on atomic numbers rather than atomic weights. Most of the features of this table are the same as Mendeleev's; the most important difference is that subgroups A and B occupy separate columns. The long form periodic table will be found following the table of contents.

The horizontal rows are called periods. The number of elements in each is indicated as follows:

| Period 1 | 2 elements | (H, He) |
|----------|------------|---------|
| Period 2 | 8 elements | (Li to Ne) |
| Period 3 | 8 elements | (Na to Ar) |
| Period 4 | 18 elements | (K to Kr) |
| Period 5 | 18 elements | (Rb to Xe) |
| Period 6 | 32 elements | (Cs to Rn) |
| Period 7 | 17 elements | (Fr to Lw) |

The vertical columns are called groups. The 16 groups are IA to VIIA, IB to VIIB, VIII and group O.

Elements of the same group, or family, have similar properties because they have the same number of valence electrons, that is, they have the same number of electrons in the outermost shell. Thus, lithium, sodium, potassium, rubidium, and cesium of the alkali family (group IA) all have one valence electron; fluorine, chlorine, bromine, and iodine of the halogen family (group VIIA) all have seven valence electrons.

The light metals are located in the left portion of the table (IA, IIA). The heavy metals, or transition elements, are located in the central portion of the table (IIIB to VIIB, VIII, IB, IIB). The nonmetals are located in the upper right corner of the table (IIIA to VIIA).

The noble gases are located on the extreme right under group O.

The lanthanide series (or rare earths) and the actinide series, both of which have almost identical properties, belong to group IIIB.

## VALUE OF THE PERIODIC TABLE

The periodic table is of value in assisting the memory. Because different elements in the same family have similar properties, it is not necessary to learn the properties of each individual element. Thus, the periodic table simplifies the study of chemistry.

The periodic table has been used to predict new elements. At the time Mendeleev made his periodic arrangement, there were several blank spaces in the table. He assumed that the missing elements would have properties close to the average of those of their neighbors; thus he was able to predict their properties before the elements themselves were discovered. Below, some of the properties predicted for gallium are compared with the actual properties of the element.

|  | Predicted Properties | Actual Properties |
|---|---|---|
| Atomic weight | 69 | 69.75 |
| Melting point | low | 30.2°C. |
| Specific gravity | 5.9 | 5.95 |
| Formula of oxide | $X_2O_3$ | $Ga_2O_3$ |

Another use of the periodic table is in correcting errors. When the value of a property of an element deviates too much from those of the neighboring elements, an error can be suspected. Repeated determinations may yield the correct value. For instance, the density of cesium was first reported as 2.4, which, according to the periodic table, was too high. Later determination gave the value of 1.9.

# 4. NUCLEAR CHEMISTRY AND ATOMIC ENERGY

In ordinary chemical reactions the changes involved are in the valence electrons outside the nuclei of the atoms. In nuclear reactions, such as in radioactivity, changes take place in the atomic nuclei, resulting in the formation of new elements. The change of one element into another is called *transmutation.*

*Radioactivity* is a process in which the nuclei of certain heavy atoms disintegrate by emitting different forms of radiation. The discovery of natural disintegration of such radioactive elements as uranium and radium led to the development of nuclear chemistry and atomic energy.

The following scientists made significant contributions to the field of nuclear chemistry.

In 1895, Wilhelm Roentgen, a German physicist, discovered X rays and found that they can penetrate opaque objects.

In 1896, A. H. Becquerel, a French physicist, found that uranium compounds emit rays that are able to darken a photographic plate in the absence of light.

In 1898, Marie and Pierre Curie of France, by a careful study of uranium ores, determined that thorium was a radioactive element and discovered polonium and radium. They found that radium possesses radioactivity about three million times that of uranium.

Ernest Rutherford, a New Zealand physicist, established the identity of alpha ($\alpha$), beta ($\beta$), and gamma ($\gamma$) rays of radioactive elements, and in 1919 produced the first artificial transmutation of an element (that of nitrogen into oxygen).

In 1939, Otto Hahn and F. Strassmann in Germany discovered nuclear fission, the basis of the atomic bomb and of atomic energy.

In 1942, Enrico Fermi, an Italian-American physicist, designed the first atomic piles and produced the first chain reaction.

## NATURAL RADIOACTIVITY

There are three types of natural radiation: alpha, beta, and gamma rays. These rays are emitted when natural radioactive elements such as uranium and thorium disintegrate.

*Alpha rays* consist of positively charged alpha particles which are composed of two protons and two neutrons. They are, therefore, the nuclei of helium atoms. Their velocity is over 10,000 miles per second, but their penetrating power is slight. They produce ionic particles in other materials by knocking off electrons from them.

*Beta rays* are made up of beta particles which are rapidly moving streams of electrons and, therefore, are negatively charged. They have an initial velocity of 100,000 miles per second and can penetrate thin sheets of metal. They possess ionizing properties less powerful than those of alpha particles because of their much smaller mass.

*Gamma rays* are similar to but more penetrating than X rays, which are electromagnetic radiation of extremely short wavelength. They also possess ionizing properties. They travel with the velocity of light and carry no electric charges. They can penetrate thick layers of metal.

**Radioactive Disintegration Series.** There are four known radioactive disintegration series. The uranium series is the most important because it is the way the elements are transmuted in nature. For example, U-238, by giving off an alpha particle, disintegrates to Th-234; Th-234, by giving off a beta particle, disintegrates to Pa-234, and so on. The uranium disintegration series is shown below.

$$^{238}_{92}\text{U} \xrightarrow[\alpha]{4.5 \times 10^9 \text{y}} {}^{234}_{90}\text{Th} \xrightarrow[\beta]{24.6\text{d}} {}^{234}_{91}\text{Pa} \xrightarrow[\beta]{1.14\text{m}} {}^{234}_{92}\text{U} \xrightarrow[\alpha]{2.35 \times 10^5 \text{y}}$$

$$^{230}_{90}\text{Th} \xrightarrow[\alpha]{8.0 \times 10^4 \text{y}} {}^{226}_{88}\text{Ra} \xrightarrow[\alpha]{1.6 \times 10^3 \text{y}} {}^{222}_{86}\text{Rn} \xrightarrow[\alpha]{3.82\text{d}} {}^{218}_{84}\text{Po} \xrightarrow[\alpha]{3\text{m}}$$

$$^{214}_{82}\text{Pb} \xrightarrow[\beta]{26.8\text{m}} {}^{214}_{83}\text{Bi} \xrightarrow[\beta]{19.7\text{m}} {}^{214}_{84}\text{Po} \xrightarrow[\alpha]{1.5 \times 10^{-4}\text{s}} {}^{210}_{82}\text{Pb} \xrightarrow[\beta]{22\text{y}}$$

$$^{210}_{83}\text{Bi} \xrightarrow[\beta]{5\text{d}} {}^{210}_{84}\text{Po} \xrightarrow[\alpha]{140\text{d}} {}^{206}_{82}\text{Pb}$$

**Half-life.** The half-life of a radioactive element is the time it takes for one-half the initial quantity of that element to decay or to transmute to other elements. The figure given above each arrow in the uranium disintegration series is the half-life of the preceding element in years, days, minutes, or seconds ( y, d, m, or s). The Greek letters under the arrow indicate the types of radiation emitted, $\alpha$ signifying alpha particles and $\beta$ signifying beta particles.

**Nuclear Equations.** In a nuclear equation the total mass and total nuclear charges are always equal on both sides of the arrow. The mass or atomic weight of the element is represented by the superscript and the atomic number or charge by the subscript.

*Example 1:*

$$_{92}^{238}\text{U} \longrightarrow {}_{90}^{234}\text{Th} + {}_{2}^{4}\text{He}$$

Uranium          Thorium          Alpha
                                   particle

*Example 2:*

$$_{90}^{234}\text{Th} \longrightarrow {}_{-1}^{0}\text{e} + {}_{91}^{234}\text{Pa}$$

Thorium          Beta          Protactinium
                 particle

The following equation shows why the discharge of an electron in the form of a beta particle from the nucleus causes the atomic number to increase by one. The total effect is that one neutron becomes a proton by losing the negative charge. Since the weight of a beta particle is negligible, the total atomic weight remains unchanged.

$$\left(\begin{array}{c} 90+ \\ 144\pm \end{array}\right) \longrightarrow \ominus + \left(\begin{array}{c} 91+ \\ 143\pm \end{array}\right)$$

Thorium                    Beta          Protactinium
nucleus                  particle          nucleus

## ARTIFICIAL NUCLEAR TRANSFORMATION

By bombarding the nuclei of ordinary elements with high-speed projectiles such as alpha particles, protons, and neutrons, it is possible to transform them into other elements, which may or may not be radioactive.

**Artificial Transmutation Forming Nonradioactive Elements.** This was achieved in 1919 by Rutherford. He found that when nitrogen was bombarded with alpha particles from radium, an isotope of oxygen was formed.

$$^{14}_7N \; + \; ^4_2He \; \rightarrow \; ^1_1H \; + \; ^{17}_8O$$

Nitrogen    Alpha    Proton    Oxygen
particles            isotope

**Artificial Radioactivity.** In 1934 Frédéric Joliot and his wife, Irene Curie, first produced artificially radioactive substances by bombarding boron, aluminum, and magnesium with alpha particles.

$$^{10}_5B \; + \; ^4_2He \; \rightarrow \; ^{13}_7N \; + \; ^1_0N$$

Boron    Alpha    Radioactive    Neutron
isotope   particles    nitrogen

$$^{13}_7N \; \rightarrow \; ^{13}_6C \; + \; ^0_{+1}e$$

Radioactive    Carbon    Positron
nitrogen     isotope

Since then, more than 500 radioactive isotopes have been produced. Many of these have found wide use in medicine and scientific research.

**Particle Accelerators.** The cyclotron and other "atom smashing" machines have been used for accelerating protons, deuterons, and alpha particles to produce high enough energy to achieve nuclear transformations. The cyclotron was invented by E. O. Lawrence at the University of California in 1929. Dr. Lawrence received the Nobel Prize for this achievement.

**Synthesis of New Elements.** By suitable nuclear reactions, scientists have made a number of new elements. They include elements 43, 61, 85, 87, and the transuranium elements 93 to 103. These elements all have extremely short half-lives.

## ATOMIC ENERGY

Atomic energy, or nuclear energy, is energy derived from the loss of mass occurring in certain types of nuclear reactions. Thus a loss of mass results in a gain of energy. This makes it necessary to combine the Law of Conservation of Mass with the Law of Conservation of Energy into a new law, the Law of Conservation of Mass-Energy, which may be stated: The total amount of mass and energy in the universe cannot be changed but is interconvertible.

In Einstein's mass-energy equivalence equation,

$$E = mc^2$$

where $E$ represents energy, $m$ mass, and $c$ the velocity of light. By the use of this equation it can be determined, for example, that the annihilation of 1 gram of matter will produce $9 \times 10^{20}$ ergs, or $2.15 \times 10^{13}$ calories, which is roughly the energy obtained by burning 3,000 tons of coal.

**Nuclear Fission.** When a uranium-235 or a plutonium-239 atom is bombarded by a slow-moving neutron, it splits into two parts of approximately equal mass with the simultaneous conversion of a small amount of mass into a tremendous amount of energy.

CHAIN REACTION. Because two or three neutrons are produced for each neutron consumed in a fission reaction, a continuous nuclear reaction, called a chain reaction, can thus be sustained.

One of the many ways in which the uranium atom can split is represented by the following equation and Fig. 4.1.

$$^{235}_{92}U + {}^{1}_{0}n \rightarrow {}^{96}_{38}Sr + {}^{138}_{54}Xe + 2{}^{1}_{0}n + 0.1\% \text{ matter converted into energy}$$

ATOMIC BOMB. The explosion of an atomic bomb is due to the sudden release of a large amount of energy when fission takes place in an uncontrolled nuclear chain reaction.

ATOMIC REACTORS. The basis of atomic reactors, or atomic piles, is also nuclear fission, except that here the chain reaction is never allowed to go out of control. Cadmium or boron rods which absorb neutrons are used in the piles. The heat liberated in atomic reactors may be used to generate electricity or to propel aircraft carriers, merchant ships, and submarines. Most of the radioactive isotopes used in industry, medicine, and scientific research are produced in atomic reactors.

**Nuclear Fusion.** Nuclei of hydrogen or its isotopes, deuterium and tritium, fuse to a helium nucleus at the high temperature of an exploding atomic bomb (10,000,000°C.). This is the basis of the hydrogen bomb. Because of greater conversion of matter into energy, more energy is released by a hydrogen bomb than by an atomic bomb.

$$^{2}_{1}H + {}^{3}_{1}H \rightarrow {}^{4}_{2}He + {}^{1}_{0}n + 0.4\% \text{ matter converted into energy}$$

Deuterium   Tritium   Helium   Neutron

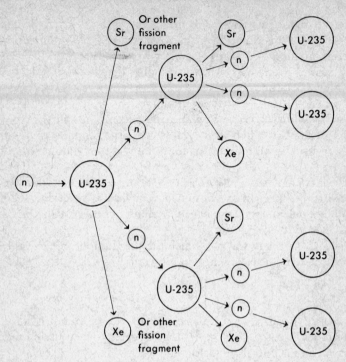

FIG. 4.1.    Fission of uranium showing chain reaction.

## DETECTION OF RADIOACTIVITY

The following instruments have been used for the detection of radiation produced by radioactive materials:

**Electroscope.**    This device consists of a rod which pierces the stopper of a glass flask.  The rod has a brass knob at its outer end and two strips of gold leaf at the other.  When the electroscope is charged by rubbing the knob with cat's fur, the gold leaves separate since they bear similar charges.  Ionized air molecules produced by radioactivity will neutralize the charge and cause the gold leaves to come together.

**Wilson Cloud Chamber.**    This instrument makes visible the paths of charged particles through supersaturated vapor.  Liquid droplets condense on the ions formed in the passage of the particle.  The fog tracks may be observed visually or photographed.

**Geiger-Müller Counter.**    The Geiger-Müller counter consists of a metallic tube filled with gas molecules under reduced pressure. A thin wire, insulated from the tube, runs down the middle. These two metals are connected to a battery. Radiation entering the tube, by ionizing the gas molecules, triggers a momentary discharge from a high-voltage circuit. The resulting surge is detected as a click in the earphone, a flash in the neon bulb, or by a reading on the meter.

**Film Badge.** Radiation is indicated by the darkening of the X-ray film contained inside the badge.

## RADIATION ILLNESS AND DAMAGE

When radiations penetrate living cells, they leave in their tracks a mixture of strange, unstable ions. As these seek stability, bonds within them may break permanently. The ions may combine with neighbors to form new molecules which are foreign to the cell; the result is radiation illness. The seriousness of radiation illness and the extent of damage depend on the type of radiation and the exposure dosage.

The principal symptoms of radiation illness are nausea, vomiting, diarrhea, internal hemorrhage, and a feeling of weakness. Radiation damage includes the following:

1. Severe burns of the skin and loss of hair.
2. Destructive effect on bone marrow and reduction of red cells, causing anemia.
3. Damage to genes and chromosomes, producing mutative changes.
4. Drop in the level of white cells.
5. Damage to enzymes, impairing cell metabolism.

## USES OF RADIOISOTOPES IN MEDICINE

*Radioactive sodium,* Na-24, when incorporated into sodium chloride and introduced into the bloodstream, helps to pinpoint the location of a blood clot caused by gangrene and thus to determine the exact site of amputation.

*Radioactive phosphorus,* P-32, is used to pinpoint brain tumors because tumors absorb phosphorus faster than do healthy brain cells. It is also used in the treatment of certain types of leukemia.

*Radioactive iodine,* I-131, is used to test thyroid function. It is also used to treat cancer of the thyroid and hyperthyroidism or Graves' disease.

*Radioactive cobalt,* Co-60, is used in the treatment of cancer, since it destroys cancerous tissue more rapidly than normal tissue. Co-60 has many advantages over radium and radon, which were formerly used for cancer treatment.

## USES OF RADIOISOTOPES IN RESEARCH

Scientists use radioisotopes as *tracers* to determine how chemicals act in the bodies of plants and animals. Some important research using radioisotope tracers is given below.

1. The complex reactions of photosynthesis have been determined by exposing green plants to carbon dioxide containing carbon-14 and analyzing the products for radioactive carbon.

2. Radioactive phosphorus in phosphate fertilizer has been used as a rapid and accurate means of assessing the rate of fertilizer incorporation in grown plants.

3. The biosynthesis of cholesterol from acetate has been demonstrated by feeding laboratory animals acetate tagged with radioactive carbon and finding radioactivity in the cholesterol molecule.

4. By feeding the subject table salt containing radioactive sodium and detecting radioactivity from his hand, it is possible to determine how much time has elapsed between the eating of the table salt and its first appearance in the bloodstream.

5. After the intravenous injection of radioactive iron salt, it is possible to measure with a radiation detector the rate at which iron is entering and leaving the plasma of the patient.

# 5. CHEMICAL NOTATION, VALENCE, TYPES OF CHEMICAL REACTIONS

In this chapter are discussed the basic tools of chemistry that the chemist uses to express a vast amount of information in a simple and concise way.

## CHEMICAL NOTATION

In order to simplify the description of chemical reactions, chemists use abbreviations for elements (symbols) and compounds (formulas) to indicate the amounts as well as the kinds of substances used and formed and to summarize the reaction in a chemical equation by means of these abbreviations.

**Symbols.** A chemical symbol is the first letter, or the first and another letter, of the English or Latin name of the element.

| | | | |
|---|---|---|---|
| Hydrogen | H | Potassium (Kalium) | K |
| Helium | He | Sodium (Natrium) | Na |
| Hafnium | Hf | Lead (Plumbum) | Pb |

The symbol represents one atom of the element. It also stands for one atomic weight of the element. For example, C means one atom of carbon or 12 grams of carbon. When it is desired to represent more than one atom of an element, the proper numeral is placed before the symbol, as 2C, 3C.

**Formulas.** A chemical formula represents a molecule of a compound and is made up of the symbols of the elements of which it is composed. For example, the formula for calcium chloride is $CaCl_2$. It indicates that the molecule is composed of 1 atom of calcium and 2 atoms of chlorine. When a numeral is placed before the formula, it indicates the number of molecules. Thus $2CaCl_2$ means 2 molecules of calcium chloride, and in like manner $3NaOH$ means 3 molecules of sodium hydroxide.

**Equations.** A chemical equation is a qualitative and quantitative expression of the reacting substances and their products in a chemical reaction. It indicates:

1. The nature of the reacting substances and the nature of the products.

2. The number of molecules of the reacting substances and the number of molecules of the products.

3. The relative weights of the reacting substances and the products.

4. The relative volume of the reacting substances and the products if they are gases.

Thus, the equation

$$CH_4 \ + \ 2O_2 \ \rightarrow \ CO_2 \ + \ 2H_2O$$

Methane    Oxygen    Carbon    Water
dioxide

shows (1) that methane reacts with oxygen to form carbon dioxide and water, (2) that one molecule of methane reacts with two molecules of oxygen to form one molecule of carbon dioxide and two molecules of water, (3) that 16 grams (or any other weight unit) of methane reacts with 64 grams of oxygen to form 44 grams of carbon dioxide and 36 grams of water, and (4) that one volume of methane reacts with two volumes of oxygen to form one volume of carbon dioxide and two volumes of water if the temperature is high enough so that the water is in the form of steam.

The following problems involve the weights and volumes of reactants and products. In volume problems you will have to know that one mole (the molecular weight in grams) of any gas under standard conditions of temperature (0°C.) and pressure (760 mm.) occupies approximately 22.4 liters.

*Problem 1:*   How many grams of potassium chlorate must be decomposed to form 12 grams of oxygen?

$$x \text{ gm.} \qquad\qquad\qquad 12 \text{ gm.}$$
$$2KClO_3 \qquad\qquad \rightarrow 2KCl \ + \qquad 3O_2$$
$$2[39 + 35.5 + (3 \times 16)] \qquad\qquad 3(2 \times 16)$$
$$245 \text{ gm.} \qquad\qquad\qquad 96 \text{ gm.}$$

$$\frac{x \text{ gm.}}{245 \text{ gm.}} = \frac{12 \text{ gm.}}{96 \text{ gm.}}$$

$$x = \frac{12 \times 245}{96} = 30.6 \text{ gm.}$$

*Problem 2:*   What volume of oxygen under standard conditions will be formed by decomposing 49 gm. of potassium chlorate?

$$49 \text{ gm.} \qquad\qquad\qquad x \text{ l.}$$
$$2KClO_3 \ \rightarrow \ 2KCl \ + \qquad 3O_2$$
$$245 \text{ gm.} \qquad\qquad\qquad 3 \times 22.4$$
$$67.2 \text{ l.}$$

$$\frac{49 \text{ gm.}}{245 \text{ gm.}} = \frac{x \text{ l.}}{67.2 \text{ l.}}$$

$$x = \frac{49 \times 67.2}{245} = 13.4 \text{ l.}$$

*Problem 3:* How many liters of oxygen are required to burn 10 liters of acetylene ($C_2H_2$)?

$$\begin{array}{ccc} 10 \text{ l.} & & x \text{ l.} \\ 2C_2H_2 & + & 5O_2 & \rightarrow 4CO_2 + 2H_2O \\ 2 \times 22.4 \text{ l.} & & 5 \times 22.4 \text{ l.} \end{array}$$

$$\frac{10 \text{ l.}}{2 \times 22.4 \text{ l.}} = \frac{x \text{ l.}}{5 \times 22.4 \text{ l.}}$$

$$\frac{10}{2} = \frac{x}{5}$$

$$x = \frac{5 \times 10}{2} = 25 \text{ l.}$$

BALANCING EQUATIONS. To fulfill the Law of Conservation of Mass, a chemical equation must be balanced, that is, there must be an equal number of atoms of the same element on both sides of the arrow. Generally two steps are involved.

1. Write the skeleton (unbalanced) equation for the reaction by showing both the substances reacting (reactants) and the substances produced (products). Thus, for the decomposition of potassium chlorate to potassium chloride and oxygen, we write:

$$KClO_3 \rightarrow KCl + O_2$$

2. Balance the skeleton equation by placing appropriate numbers (coefficients) in front of the formulas where they are needed until the equation is balanced. These are 2 for $KClO_3$, 2 for $KCl$, and 3 for $O_2$. Thus:

$$2KClO_3 \rightarrow 2KCl + 3O_2$$

## VALENCE

Valence is the combining capacity of an element which represents the number of atoms of hydrogen (or its equivalent) that its atom can hold in combination or displace. It also represents the number of electrons gained, lost, or shared by an atom in a chemical reaction.

Valence electrons are the electrons present in the outermost shell of an atom, since their loss, gain, or sharing determines the valence of the element.

*Electrovalence*, or ionic valence, is the valence resulting from the transfer of electrons and is equal to the number of electrons gained or lost by an element. When electrons are lost, the resulting valence is positive; when electrons are gained, the resulting valence is negative.

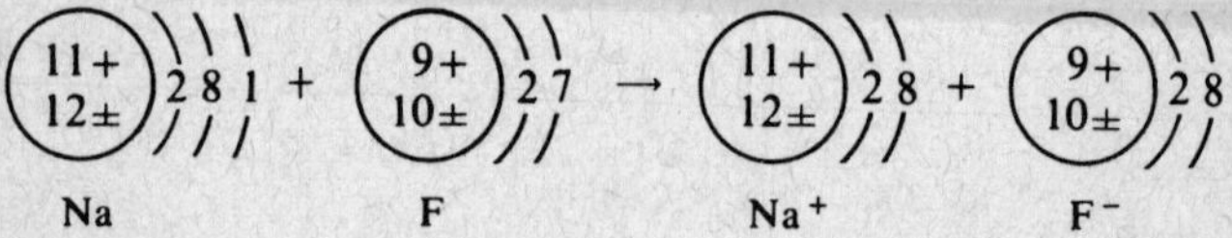

*Covalence* is the valence resulting from the sharing of electrons by combining atoms in forming a molecule. Each pair of electrons shared gives the atom a valence of 1 and is represented by one bond in the structural formula.

*Electron dot formula:*

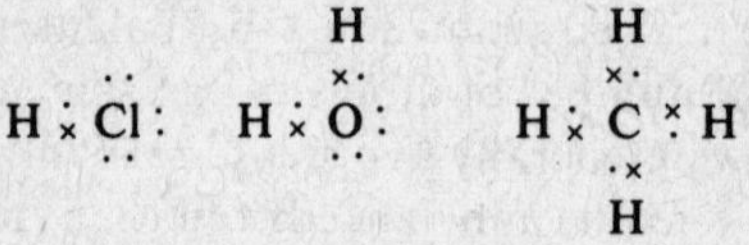

*Structural formula:*

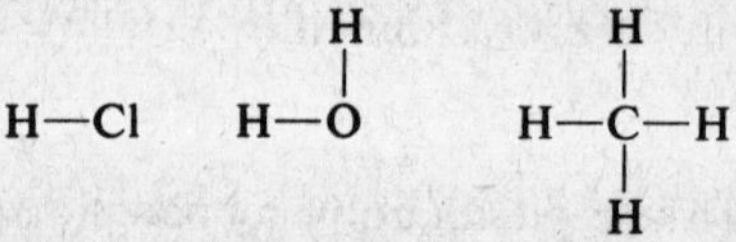

The above formulas indicate that hydrogen and chlorine each have a valence of 1, oxygen a valence of 2, and carbon a valence of 4.

Besides single atoms carrying positive or negative charges, certain groups of atoms acting as a unit can carry charges and have valence. Such groups of atoms are called radicals. Some of the common radicals are the ammonium radical $NH_4^+$, nitrate radical $NO_3^-$, sulfate radical $SO_4^{-2}$, and phosphate radical $PO_4^{-3}$.

The valence of some common elements and radicals is given in Table 5.1.

**Use of Valence in Writing Formulas.**   Since a compound as a whole is electrically neutral, the sum of the valences of the positive elements or radicals must equal the sum of the valences of the

### TABLE 5.1.  VALENCE OF SOME COMMON ELEMENTS AND RADICALS

| +1 GROUP | | −1 GROUP | |
|---|---|---|---|
| $H^+$ | hydrogen ion | $F^-$ | fluoride ion |
| $Na^+$ | sodium ion | $Cl^-$ | chloride ion |
| $K^+$ | potassium ion | $Br^-$ | bromide ion |
| $Ag^+$ | silver ion | $I^-$ | iodide ion |
| $Cu^+$ | cuprous ion | $OH^-$ | hydroxide ion |
| $Hg^+$ | mercurous ion | $NO_3^-$ | nitrate ion |
| $NH_4^+$ | ammonium ion | $NO_2^-$ | nitrite ion |
| | | $HCO_3^-$ | bicarbonate |

| +2 GROUP | | −2 GROUP | |
|---|---|---|---|
| $Ca^{+2}$ | calcium ion | $O^{-2}$ | oxide ion |
| $Mg^{+2}$ | magnesium ion | $S^{-2}$ | sulfide ion |
| $Ba^{+2}$ | barium ion | $SO_3^{-2}$ | sulfite ion |
| $Cu^{+2}$ | cupric ion | $SO_4^{-2}$ | sulfate ion |
| $Zn^{+2}$ | zinc ion | $CO_3^{-2}$ | carbonate ion |
| $Hg^{+2}$ | mercuric ion | | |
| $Fe^{+2}$ | ferrous ion | | |

| +3 GROUP | | −3 GROUP | |
|---|---|---|---|
| $Al^{+3}$ | aluminum ion | $PO_4^{-3}$ | phosphate ion |
| $Fe^{+3}$ | ferric ion | | |

negative elements or radicals.  This is illustrated in the following examples:

| | |
|---|---|
| $H^+Cl^-$ | Hydrogen chloride |
| $(NH_4^+)_2SO_4^{-2}$ | Ammonium sulfate |
| $Al^{+3}(NO_3^-)_3$ | Aluminum nitrate |

In ammonium sulfate above, we have 2 ammonium radicals each having a valence of 1-positive and 1 sulfate radical having a valence of 2-negative.  The total positive (2) equals the total negative (2).  Likewise, in aluminum nitrate, we have 1 aluminum having a valence of 3-positive and 3 nitrate radicals each having a valence of 1-negative.  Again the total positive (3) equals the total negative (3).

To write the chemical formula of a compound, it is necessary to know the valences of the elements or radicals composing the compound.  For example, to figure out the correct formula of calcium phosphate, first write down the skeleton formula with the valences of calcium and the phosphate radical:

$$Ca^{+2} \qquad PO_4^{-3}$$

Then determine the least common multiple of the opposite charges, which in this case is 6; divide this number by the number of positive charges (2) to find the correct number of calcium atoms in the formula (3), and by the number of negative charges (3) to find the correct number of the phosphate radical (2). The correct formula of calcium phosphate is therefore $Ca_3(PO_4)_2$.

The chemical formula also represents one molecular weight of the compound, and one molecular volume of the compound if it is a gas. The *molecular weight* of a compound represents the sum total of the atomic weights of all the atoms shown in the formula of the compound. For example, the molecular weight of $CaCl_2$ is $40 + (2 \times 35.5) = 111$. When the molecular weight of a compound is expressed in grams, it is called *gram-molecular weight*.

In one molecular weight of any substance, whether solid, liquid, or gas, there are $6.023 \times 10^{23}$ molecules. This number is called *Avogadro's number*.

## TYPES OF CHEMICAL REACTIONS

There are four types of chemical reactions: combination, decomposition, replacement, and double decomposition. A reversible reaction is a special kind of double decomposition.

**Combination or Synthesis.** Two or more substances react to form a more complex substance.

$$2H_2 + O_2 \rightarrow 2H_2O$$

**Decomposition.** A substance is broken down into two or more simpler substances.

$$2KClO_3 \rightarrow 2KCl + 3O_2$$

**Replacement or Displacement.** An element in a chemical compound is replaced by another element. In the following example, zinc replaces hydrogen from hydrochloric acid.

$$Zn + 2HCl \rightarrow ZnCl_2 + H_2$$

**Double Decomposition or Metathesis.** The components of two compounds exchange partners to form two different compounds.

$$BaCl_2 + H_2SO_4 \rightarrow BaSO_4 + 2HCl$$

**Reversible Reaction.** In a reversible reaction, the products that are produced in turn react to form the original reactants. For example, hydrogen and iodine readily react to produce hydrogen iodide.

$$H_2 + I_2 \rightarrow 2HI$$

The hydrogen iodide produced in turn decomposes to form hydrogen and iodine.

$$2HI \rightarrow H_2 + I_2$$

This type of two-way reaction, forward and reverse, is called a reversible reaction. It is usually represented by means of a double arrow.

$$H_2 + I_2 \rightleftharpoons 2HI$$

When the rates of the forward reaction and the reverse reaction are equal, the reaction is said to have reached *equilibrium*.

# 6. OXYGEN

Although Scheele, a Swedish chemist, prepared oxygen in 1771, the discovery of oxygen is usually credited to Joseph Priestley, an Englishman, who prepared it in 1774, because he published his results before Scheele did. But it was the great French chemist Lavoisier who recognized the significance of Priestley's discovery and named the element *oxygen*. He also demonstrated that oxygen was the active fraction of the air involved in ordinary combustion and similar processes such as corrosion, decay, and metabolism. This marked the end of the Phlogiston Theory and set the stage for a better understanding of chemical changes. For this reason, Lavoisier has been called the father of modern chemistry.

**Occurrence.** Oxygen is the most abundant of all the elements. In its free state it makes up approximately one-fifth of the air by volume. In chemical combination with other substances, oxygen constitutes by weight approximately 50 percent of the earth's crust, eight-ninths of water, from 50 to 70 percent of plant and animal tissues, and about 65 percent of the human body.

**Preparation.** Oxygen may be prepared commercially and in the laboratory by the following methods:

COMMERCIAL METHODS. Oxygen is usually prepared commercially by liquefying air by subjecting it to high pressure and low temperature and then distilling off the more volatile nitrogen. The liquid oxygen is then vaporized and compressed into steel cylinders. Another method of preparing oxygen commercially is by the electrolysis of water. An electric current is passed through water containing a little sulfuric acid to make it more conductive. Oxygen gas collects at the anode (positive pole) and hydrogen gas at the cathode (negative pole).

$$2H_2O \rightarrow 2H_2\uparrow + O_2\uparrow$$

LABORATORY METHODS. In the laboratory oxygen is easily prepared by heating potassium chlorate with a small amount of manganese dioxide. The manganese dioxide serves as a *catalyst*, which is a substance that influences the speed of a chemical reaction without being changed itself.

$$2KClO_3 \xrightarrow[\text{heat}]{\text{MnO}_2} 2KCl + 3O_2$$

Another laboratory method of preparing oxygen is Priestley's method of heating mercuric oxide.

$$2HgO \longrightarrow 2Hg + O_2\uparrow$$

Mercuric oxide    Mercury    Oxygen

**Physical Properties.** Oxygen is a colorless, odorless, and tasteless gas, slightly heavier than air. It is slightly soluble in water, 3 ml. in 100 ml. of water at ordinary temperatures. This slight solubility is important to marine vegetation and animal life and for the oxidation of decaying organic matter.

Oxygen can be liquefied by cooling it to $-118°C$. under 50 atmospheres pressure. Liquid oxygen is pale blue in color and boils at $-183°C$. at atmospheric pressure.

**Chemical Properties.** The outstanding chemical property of oxygen is its ability to combine with other elements, with moderate activity at ordinary temperatures and more vigorously at higher temperatures. The union of a substance with oxygen is called *oxidation*. Rapid oxidation that is accompanied by heat and light, such as ordinary burning, is called *combustion*.

Oxygen reacts with the more active metals to form metallic oxides.

$$4Fe + 3O_2 \longrightarrow 2Fe_2O_3$$

Iron    Oxygen    Ferric
oxide

Oxygen reacts with nonmetals to form nonmetallic oxides.

$$C + O_2 \longrightarrow CO_2$$

Carbon    Oxygen    Carbon
dioxide

Oxygen also reacts with many organic compounds.

$$C_6H_{12}O_6 + 6O_2 \longrightarrow 6CO_2 + 6H_2O$$

Glucose    Oxygen    Carbon    Water
dioxide

SLOW OXIDATION. Slow oxidation is oxidation not accompanied by noticeable light and heat. The corrosion of metals and the decaying of organic matter are examples of slow oxidation.

*Corrosion of Metals.* Active and moderately active metals when exposed to air and moisture combine with oxygen to form oxides. The rusting of iron is a familiar example.

*Decay of Organic Matter.* Decay is the slow oxidation of organic matter in the presence of bacteria and moisture. With sufficient oxygen, decaying substances are rendered harmless by this natural process.

*Spontaneous Combustion.* When slow oxidation takes place and the heat produced cannot escape because of poor air circulation, the temperature may continue to rise until the kindling temperature is reached and the material (oily rags, moist hay, etc.) bursts into flame.

EXTINGUISHING FIRES. In order for combustion to take place, it is necessary to raise the temperature of the substance to its *kindling temperature.* This is the lowest temperature at which a substance will burst into flame. Based on this principle, fires are extinguished in two ways: (1) by cooling the burning material to below its kindling temperature, as by water, and (2) by excluding oxygen from the burning material, as by carbon dioxide or sand.

There are a number of types of fire extinguishers. In the soda-acid type, carbon dioxide gas is generated when the extinguisher is inverted and sulfuric acid from a small bottle pours into a sodium bicarbonate solution.

$$H_2SO_4 + NaHCO_3 \rightarrow NaHSO_4 + H_2O + CO_2$$

The carbon dioxide gas, being heavier than air, covers the burning material and excludes oxygen from it.

Another type of extinguisher contains liquid carbon dioxide under pressure. When pressure is released, the liquid is vaporized to carbon dioxide gas which can be directed on the fire through a nozzle. The advantages of this type over the soda-acid extinguisher are that it has a greater capacity, causes less damage, and can be used safely on many types of fires.

Foamite extinguishers contain solutions of sodium bicarbonate, aluminum sulfate, and licorice extract, which helps to form a stable carbon dioxide foam. This type of extinguisher is especially effective on gasoline or oil fires, on which water cannot be used.

"Pyrene" extinguishers contain carbon tetrachloride, which extinguishes fire by blanketing the burning material with its heavy vapor. Because carbon tetrachloride produces toxic fumes, care must be exercised in the use of this type of extinguisher.

**Uses.** Oxygen supports life. In the higher animals the blood carries it from the lungs to the tissues where it oxidizes foodstuffs, thus providing the body with heat and energy. It is essential in the combustion of fuels, such as coal, natural gas, and gasoline.

Compressed oxygen in tanks is used to relieve suffocation in cases of pneumonia or gaseous poisoning; in the operating room in connection with anesthetics; and in cardiac failure. It is standard equipment for aviators, miners, mountain climbers, and deep-sea divers. Oxygen is a good disinfecting agent and a natural water purifier. Its most important industrial use is in the oxyacetylene or oxyhydrogen torch for cutting and welding metals.

**Ozone.** Ozone is a special form of oxygen containing three atoms of oxygen per molecule instead of two. This property of certain elements to exist in more than one molecular form and therefore to have somewhat different properties is called *allotropism.*

Ozone is a blue gas with a pungent odor. Small amounts of ozone are produced by lightning and by electric motors. Commercially it is produced by passing air between two plates which are charged at a high potential of several thousand volts of alternating current.

Ozone is more active than oxygen because of the *nascent oxygen* that is released when ozone decomposes. Nascent oxygen is atomic oxygen, which is more active chemically than ordinary oxygen.

$$O_3 \rightarrow O_2 + O$$

| Ozone | Ordinary oxygen | Nascent oxygen |
|-------|-----------------|----------------|

Ozone is used for deodorizing, disinfecting, and bleaching purposes.

# 7. OXIDATION AND REDUCTION

Oxidation and reduction may be defined in three different ways.

1. Oxidation is a chemical reaction in which an element combines with oxygen or any other nonmetal to form a compound. Reduction is the taking away of oxygen or any other nonmetal from a compound by another element. In the reactions below, the carbon, hydrogen, and iron are oxidized.

$$C + O_2 \rightarrow CO_2$$

$$H_2 + Cl_2 \rightarrow 2HCl$$

$$Fe + S \rightarrow FeS$$

In the following reaction, the element, copper, of the cupric oxide is reduced.

$$CuO + H_2 \rightarrow Cu + H_2O$$

2. Oxidation involves an increase in positive valence or a decrease in negative valence. Reduction is just the opposite.

$$SnCl_2 + 2FeCl_3 \rightarrow SnCl_4 + 2FeCl_2$$

$$(Sn^{+2} + 2Fe^{+3} \rightarrow Sn^{+4} + 2Fe^{+2})$$

Tin is oxidized from +2 to +4, while iron is reduced from +3 to +2.

$$2NaBr + Cl_2 \rightarrow 2NaCl + Br_2$$

$$(2Br^- + Cl_2^0 \rightarrow Br_2^0 + 2Cl^-)$$

Bromine is oxidized from −1 to 0, while chlorine is reduced from 0 to −1.

3. Oxidation is the loss of one or more electrons. Reduction is the gain of one or more electrons.

$$3Na + AlCl_3 \rightarrow 3NaCl + Al$$

$$(Na^0 - e^- \rightarrow Na^+; \; Al_3^+ + 3e^- \rightarrow Al^0)$$

Sodium is oxidized since it loses an electron; aluminum is reduced since it gains three electrons.

In a chemical reaction, oxidation and reduction occur simultaneously and to the same extent. The electrons lost by one element must be gained by another.

$$Zn + 2HCl \rightarrow ZnCl_2 + H_2$$
$$(Zn^0 + 2H^+ \rightarrow Zn^{+2} + H_2^0)$$

Zinc is oxidized while hydrogen is reduced. The zinc atom loses two electrons to become a zinc ion, while the two hydrogen ions gain two electrons to become elemental hydrogen.

**Oxidizing Agents and Reducing Agents.**   The agent that causes the oxidation of an element or compound is called an oxidizing agent, while the agent that causes the reduction of an element or compound is called a reducing agent.

$$Mg + Br_2 \rightarrow MgBr_2$$

Magnesium is oxidized; therefore it is the reducing agent.   Bromine is reduced; therefore it is the oxidizing agent.

**Uses of Oxidizing and Reducing Agents.**   Oxidizing and reducing agents find wide application as antiseptic and stain-removing agents.

ANTISEPTIC AGENTS.   Most of the common antiseptics owe their efficacy to the bacteria-destroying property of the nascent oxygen that is released.   They include such oxidizing agents as hydrogen peroxide ($H_2O_2$) and potassium permanganate ($KMnO_4$).

STAIN REMOVAL.   Certain pigments and stains are converted to colorless compounds by oxidizing agents, while others are decolorized by reducing agents.   Hydrogen peroxide, an oxidizing agent, is used for removing blood stains and for bleaching organic material.   Javell water and household bleach, also oxidizing agents, are solutions of sodium hypochlorite and are used for bleaching cotton goods.

The reducing agents include oxalic acid, sodium thiosulfate, and sulfur dioxide gas.   Oxalic acid $(COOH)_2$ is used for removing iron stains, sodium thiosulfate ($Na_2S_2O_3$) for removing iodine stains, and sulfur dioxide ($SO_2$) for bleaching straw, silk, wool, and fruits.

# 8. WATER AND HYDROGEN PEROXIDE

Water and hydrogen peroxide are compounds containing hydrogen and oxygen. While water is a very stable compound, hydrogen peroxide is extremely unstable.

## WATER

Water is a chemical compound made up of two atoms of hydrogen and one atom of oxygen. Lord Cavendish in 1781 synthesized water from its elements by exploding a mixture of hydrogen and oxygen gases. He also established the two-to-one ratio by volume of these two elements in the compound. The ratio by weight is one part of hydrogen to eight parts of oxygen.

**Occurrence.** Water is the most abundant compound in nature. About three-fourths of the surface of the earth is covered with water. Soil contains enormous amounts. In the atmosphere, large quantities of water are present in the form of dew, fog, rain, and snow. All animal and plant tissues contain water; the human body is 60 percent water. Foods contain 10 to 90 percent.

**Physical Properties.** Pure water is a colorless, odorless, tasteless liquid. It freezes at 0°C. (32°F.) and boils at 100°C. (212°F.) at 760 mm. pressure.

Water reaches its maximum density at 4°C. At this temperature 1 ml. of water weighs 1 gram. The density of water is used as a standard and is designated as 1. Thus the *specific gravity* of a substance is its density in grams per milliliter compared to the density of water. When water turns to ice, it expands. This unusual phenomenon causes water pipes to break in cold weather. Because water becomes less dense when it expands, ice floats on the surface of the water. If it did not, the entire body of water would freeze solid, making aquatic life impossible.

Water has a high *specific heat* value of 1, since it takes 1 calorie of heat to raise the temperature of 1 gram of water 1°C. Water also has a high *heat of fusion*. It takes 80 calories of heat to change 1 gram of ice at 0°C. into water at the same temperature.

Its *heat of vaporization* is also high. It takes 540 calories of heat to change 1 gram of water at 100°C. into steam at 100°C.

Water is commonly known as the universal solvent, because more substances will dissolve in it than in any other liquid.

**Chemical Properties.** Water is very stable. It may be heated to 2,000°C. without appreciable decomposition.

Water reacts with active metals like sodium, potassium, and calcium to liberate hydrogen and form a base.

$$2Na \ + \ 2H_2O \rightarrow \ H_2 \ + \ 2NaOH$$

Sodium    Water    Hydrogen    Sodium hydroxide

Water reacts with metallic oxides to form bases.

$$CaO \ + \ H_2O \rightarrow Ca(OH)_2$$

Calcium oxide    Water    Calcium hydroxide

Water reacts with nonmetallic oxides to form acids.

$$SO_3 \ + \ H_2O \rightarrow H_2SO_4$$

Sulfur trioxide    Water    Sulfuric acid

HYDROLYSIS. Hydrolysis is a double decomposition reaction in which water is one of the reacting substances. Each of the resulting compounds contains part of the water.

Water will hydrolyze certain salts to form an acid and a base.

$$NH_4Cl \ + \ HOH \rightarrow \ HCl \ + \ NH_4OH$$

Ammonium chloride    Water    Hydrochloric acid    Ammonium hydroxide

Water will also hydrolyze certain complex organic compounds to form simpler compounds. This is the basis of the digestion of foods.

$$Fat \ + \ Water \rightarrow Fatty \ acid \ + \ Glycerol$$

HYDRATES. Water combines with certain salts to form hydrates. The water held in combination is called the *water of crystallization* or the *water of hydration.*

$$CuSO_4 \ + \ 5H_2O \rightarrow CuSO_4 \cdot 5H_2O$$

Anhydrous copper sulfate    Water    Copper sulfate pentahydrate

When a hydrate loses its water of crystallization at room temperature, it is said to be *efflorescent*. An anhydrous salt that has the tendency to absorb moisture from the air at room temperature is said to be *hygroscopic*. If the salt absorbs sufficient moisture to become a liquid, it is called *deliquescent*. The use of anhydrous calcium chloride crystals on country roads to keep down the dust in dry weather is based on this property.

A completely or partially dehydrated salt often reverts back to the hydrate form when mixed with enough water.

$$(CaSO_4)_2 \cdot H_2O + 3H_2O \rightarrow 2CaSO_4 \cdot 2H_2O$$

Plaster of Paris    Water    Gypsum

The use of plaster of Paris in making molds, casts, stucco, and plaster wallboards is based upon this reaction.

**Purification of Water.** Natural waters containing dissolved and suspended impurities are purified by one or more of the following methods.

FILTRATION. Filtration consists in passing natural water through beds of sand and gravel to hold back suspended matter and most of the bacteria, and sometimes also through charcoal filters to remove odorous gases such as hydrogen sulfide and ammonia.

AERATION. Spraying water into the air helps to oxidize the organic matter, as well as to remove objectionable odors and tastes.

CHLORINATION AND OZONATION. Chlorine and, less frequently, ozone are used to destroy bacteria and other microorganisms.

DISTILLATION. Distillation, in which water is vaporized and then condensed, removes all nonvolatile solids. Distilled water is the purest water obtainable. It is used widely in the preparation of solutions in the laboratory and in the hospital.

BOILING. In case of emergency, water may be made safe for drinking purposes by boiling for 10 to 15 minutes, a process which does not remove any dissolved material, but does kill microorganisms that may cause disease.

**Water Softening.** Water containing calcium or magnesium ions is called *hard water*. When hard water is boiled, these salts form deposits inside boilers, pipes, and tea kettles. Calcium or magnesium ions also combine with the soluble sodium or potassium soap to form the insoluble calcium or magnesium soap which we call curd or scum.

The hardness of water can be temporary or permanent. *Temporary hard water* contains dissolved calcium or magnesium bicarbonate. It can be softened by boiling as well as by the addition of lime. In the following equation, boiling decomposes the soluble calcium bicarbonate. The arrow pointing downward after the $CaCO_3$ indicates an insoluble substance or precipitate. The triangle over the reaction arrow is a symbol for heat.

$$Ca(HCO_3)_2 \xrightarrow{\triangle} CaCO_3\downarrow + H_2O + CO_2\uparrow$$

Calcium bicarbonate    Calcium carbonate    Water    Carbon
(soluble)          (insoluble)        dioxide

The following equation shows the precipitation of calcium carbonate by the use of lime.

$$Ca(HCO_3)_2 + Ca(OH)_2 \rightarrow 2CaCO_3\downarrow + 2H_2O$$

*Permanent hard water* contains dissolved calcium or magnesium sulfate or chloride. It cannot be softened by boiling. It has to be treated with chemicals that will remove the calcium or magnesium ions. The reaction with washing soda (sodium carbonate) is as follows:

$$CaSO_4 + Na_2CO_3 \rightarrow CaCO_3\downarrow + Na_2SO_4$$

Calcium    Sodium    Calcium    Sodium
sulfate     carbonate   carbonate   sulfate

Both types of hard water may be softened by passing the water through modern water softeners, zeolite (Permutit process), or ion exchange resins, whose sodium ions are exchanged for calcium or magnesium ions.

**The Importance of Water.** Water is essential to life. About 60 percent of the human body is water. Man cannot live without water even for a few days. It is essential to the following body functions: the processes of digestion; the transport of food to the tissues; the elimination of body wastes; the circulation of body fluids (blood, lymph, and interstitial fluid); the maintenance of the electrolyte and osmotic pressure balance of the body; and the regulation of body temperature.

Water is the most important solvent and catalyst. It also has an effect on climate; large bodies of water help to moderate temperatures.

# HYDROGEN PEROXIDE

Hydrogen peroxide ($H_2O_2$) is prepared by treating a metallic peroxide with an acid.

$$BaO_2 + H_2SO_4 \rightarrow H_2O_2 + BaSO_4$$

| Barium peroxide | Sulfuric acid | Hydrogen peroxide | Barium sulfate |

It is a syrupy, colorless liquid with a characteristic odor and taste. It is quite unstable, readily liberating nascent oxygen in the presence of the organic enzyme catalase and other chemical agents.

$$H_2O_2 \rightarrow H_2O + O$$

Pharmaceutical hydrogen peroxide is a 3 percent solution to which a preservative has been added to prevent its decomposition by light.

Hydrogen peroxide is a good oxidizing agent and is widely used for bleaching and as a disinfectant and antiseptic. In dilute solution it is used as a mild bleaching agent for wool, silk, and hair. Because the oxygen liberated produces effervescence, hydrogen peroxide is used to loosen bloody bandages or pus from infected wounds. It is also sometimes used as a mouthwash to remove mucus.

# 9. SOLUTIONS AND COLLOIDAL DISPERSIONS

There are two types of solutions, true solutions and colloidal solutions or dispersions. In true solutions the dissolved particles are small molecules or ions. In colloidal dispersions the dispersed particles are aggregates of small molecules or giant molecules.

## SOLUTIONS

A solution is a homogeneous mixture whose components may be varied within certain limits. The *solute* is the substance dissolved, which may be a gas, liquid, or solid. The *solvent* is the substance in which the solute is dissolved. It is usually a liquid. When the solvent is water, the solution is called an *aqueous solution*, as, for example, salt or sugar dissolved in water. When the solvent is alcohol, the solution is called a *tincture*, such as tincture of iodine.

**Kinds of Solutions.** Because a solution is a mixture, the amount of solute it contains is variable. A *dilute solution* contains a small percentage of solute. A *concentrated solution* contains a large percentage of solute. A *saturated solution* is one in which the solvent contains as much of the solute as it can hold at that temperature and pressure. An *unsaturated solution* is one which contains less dissolved solute than a saturated solution. A *supersaturated solution* is one which contains more dissolved solute than a saturated solution at the same temperature. It is usually prepared by allowing a saturated solution prepared at a higher temperature to cool slowly without being disturbed.

**Factors Affecting Solubility.** The extent of solubility of a solute in a solvent depends on the following factors:

NATURE OF THE SOLUTE AND THE SOLVENT. As a general rule, inorganic compounds are soluble in water but insoluble in organic solvents, which readily dissolve organic compounds. Thus sodium chloride is soluble in water, and fats and oils are soluble in ether, benzene, and carbon tetrachloride.

SURFACE AREA.   A finely divided solute will dissolve more readily because more surface area is exposed to the solvent.

AGITATION.   Agitation, such as stirring, increases the rate of solution of the solute since fresh solvent continually comes in contact with the solute.

TEMPERATURE.   With a few exceptions, the solubility of solids and liquids increases with temperature (see Fig. 9.1), whereas the solubility of gases decreases with increase in temperature.

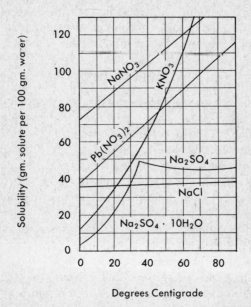

FIG. 9.1.   Solubility of several substances at various temperatures.

PRESSURE.   Pressure does not affect the solubility of solids and liquids, but an increase in pressure increases the solubility of gases.   The effect of pressure upon the solubility of a gas may be stated by Henry's Law: The weight of a gas dissolved by a given volume of a liquid is directly proportional to the pressure of the gas.

**Methods of Expressing Concentrations.**   The concentration of a solution may be expressed in a number of ways.   The following are in common use:

WEIGHT-VOLUME PERCENTAGE.   By this method, the concentration of a solution is expressed as grams of solute in 100 ml.

of solution. Thus the isotonic 0.9 percent physiological saline used in injections contains 0.9 gram of sodium chloride in 100 ml. of solution.

MOLAR SOLUTION. A molar solution is one that contains one mole (one molecular weight in grams) of the solute in one liter of solution. For example, the molecular weight of sulfuric acid is 98. A liter of solution containing 98 grams of sulfuric acid is, therefore, a one molar (1M) solution.

NORMAL SOLUTION. A normal solution is one that contains one equivalent (one equivalent weight in grams) of the solute in one liter of solution. The gram-equivalent weight is that quantity of a substance which will liberate or combine with 1.008 grams of hydrogen. For example, since sulfuric acid ($H_2SO_4$) contains two replaceable hydrogen atoms, its equivalent weight is one-half of its molecular weight. A one normal (1N) sulfuric acid solution will therefore contain 49 grams (98 ÷ 2) of sulfuric acid in one liter of solution.

**Physical Properties of Solutions.** When a nonvolatile solute is dissolved in a solvent, certain physical properties of the solvent are altered in proportion to the molecular concentration of the solution.

VAPOR PRESSURE. The vapor pressure of the solvent, which results from the tendency of its molecules to escape from its surface, is lowered by the dissolved solute, which interferes with the evaporation of the solvent. Water at 100°C. has a vapor pressure of 760 mm. One mole of a nonelectrolyte, such as sugar, alcohol, or glycerol, dissolved in 1,000 grams of water will lower its vapor pressure to about 742 mm.

BOILING POINT. The boiling point of the solvent is raised by the dissolved solute. A solution prepared by dissolving one mole of nonelectrolyte in 1,000 grams of water will boil at 100.52°C. instead of 100°C., the boiling point of pure water.

FREEZING POINT. The freezing point of the solvent is lowered by the dissolved solute. A solution prepared by dissolving one mole of nonelectrolyte in 1,000 grams of water will freeze at −1.86°C. instead of 0°C., the freezing point of pure water.

OSMOTIC PRESSURE. A solvent containing a dissolved solute will exert an osmotic pressure that is proportional to the amount of solute dissolved.

When two solutions of the same substance but of unequal concentrations and, therefore, of unequal osmotic pressure are separated by a semipermeable membrane, water will pass from the dilute solution to the more concentrated solution, resulting in a rise in its level. This process is called *osmosis*. It explains how trees draw their water from the soil.

If plain water or a solution of low osmotic pressure (*hypotonic*) is injected into the bloodstream, water will pass into the red cells, whose inside fluid has a higher osmotic pressure. This results in the swelling and final bursting of the cells (*hemolysis*).

If a solution of higher osmotic pressure (*hypertonic*) is injected, water will pass out of the red cells, causing them to shrivel (*crenation*).

Solutions that have the same osmotic pressure as that of the blood are said to be *isotonic*. Physiological salt solution (0.9%) is such a solution. Only isotonic solutions can be safely introduced into the bloodstream.

## COLLOIDAL DISPERSIONS

The term *colloid* is derived from the Greek word meaning "gluelike." In 1861, Thomas Graham, a Scottish chemist, observed that certain substances which are easy to crystallize from solution, such as sugar and salt, would pass through animal or vegetable membranes, while other substances, such as starch and gelatin, would not. He called the first group, which form true solutions, *crystalloids*, and the second group, which do not form true solutions, *colloids*.

**Types of Colloidal Systems.**    Following are eight possible types of colloidal systems with several examples of each.

| Colloidal Systems | Examples |
| --- | --- |
| Gas in liquid | Foams, whipped cream |
| Gas in solid | Pumice, meerschaum |
| Liquid in gas | Fog, mists |
| Liquid in liquid | Mayonnaise, cream |
| Liquid in solid | Opal, pearls |
| Solid in gas | Smoke, dust clouds |
| Solid in liquid | Plasma, starch paste |
| Solid in solid | Ruby glass, alloys, black diamonds |

**Solids Dispersed in Liquids.**    In a colloidal dispersion, the substance dispersed is called the *dispersed phase* and the substance in which the colloidal material is dispersed is called the *dispersion medium*. A dispersion of a solid in a liquid is called a *sol*. There are two general types of solids dispersed in liquids, lyophobic and lyophilic.

LYOPHOBIC. A *lyophobic* (or *hydrophobic* when the dispersion medium is water) colloid shows little affinity for the dispersion medium. A colloid of this type is called a *suspensoid*. Examples of lyophobic colloidal systems are arsenic sulfide dispersed in water and gold or other metals dispersed in water.

LYOPHILIC. A *lyophilic* (or *hydrophilic* when the dispersion medium is water) colloid shows great affinity for the dispersion medium. A colloid of this type is called an *emulsoid*, and when in a semisolid state, a *gel*. Examples of lyophilic colloids are gelatin desserts, custard, and pudding.

A gel is formed when colloidal particles adsorb water molecules on their surfaces as the solution cools, causing the formation of fibers and filaments which trap a large quantity of water. The result is a semisolid structure. Upon standing the gel may contract and squeeze out some of the fluid. This process is called *syneresis*. The clotting of blood involves gel formation; the passing of serum out of the clot is the process of syneresis.

**Emulsions.** An emulsion is a colloidal dispersion of one liquid in another. When two immiscible liquids, such as oil and water, are shaken, a temporary emulsion results, which upon standing separates into two distinct layers. However, if an emulsifying agent, such as egg yolk, is added to the mixture and then the mixture is shaken, a permanent emulsion is obtained. Cream, milk, and mayonnaise dressing are common examples of permanent emulsions.

The function of the emulsifying agent is to lower the surface tension of the dispersion medium (water) and to form a protective film around each droplet of the dispersed phase (oil) to prevent them from running together.

Other emulsifying agents besides egg yolk are gums, soap, lecithin, gelatin, dextrin, and starch, which are all colloids themselves. When they are employed for the purpose of forming a protective film around other colloidal particles, they are known as *protective colloids*.

**Preparation of Colloids.** There are two ways to prepare colloidal systems: by dispersion and by condensation.

DISPERSION METHODS. In these methods a mass is subdivided into colloidal-size particles.

*Mechanical Disintegration.* This includes fine grinding, especially in a colloidal mill or homogenizer. Paint pigments and face powders are prepared in this way.

*Peptization.* Disintegration is achieved by the use of a peptizing agent which can be a liquid or certain ions. An example is the peptization of agar and gelatin by water to colloidal dispersions.

*Bredig's Arc Method.* In this method, metal wires are vaporized by striking an electric arc under water. The vapor condenses to particles of colloid size.

*Emulsification.* Emulsions are prepared by emulsifying oil and water in the presence of an emulsifying agent such as egg yolk or soap.

CONDENSATION METHODS. In these methods, molecules or atoms are made to aggregate into colloidal-size particles.

*Double Decomposition.* Arsenic sulfide is prepared by reacting arsenous oxide with hydrogen sulfide.

$$As_2O_3 + 3H_2S \rightarrow As_2S_3 + 3H_2O$$

*Hydrolysis.* Ferric chloride forms a dispersion of ferric hydroxide when hydrolyzed with boiling water.

$$FeCl_3 + 3H_2O \rightarrow 3HCl + Fe(OH)_3$$

*Reduction.* Gold sols are prepared by reducing gold chloride solution with formaldehyde.

**Properties of Colloids.** Colloids have the particle size of 1 m$\mu$ to 100 m$\mu$, which is between the particle size of true solutions and suspensions. The particles in a suspension, because of their larger size, will ultimately settle out and may be readily separated from the liquid by filtration; this is not true of colloids.

Colloidal dispersions show the *Tyndall effect,* which is the reflection of a beam of light by the individual colloidal particles. They also show *Brownian movement*, which is the random movement of colloidal particles caused by uneven bombardment by solvent molecules. This effect can best be observed by means of an ultramicroscope.

Because of their large surface area, colloids have great power of taking up or holding other molecules or ions on their surfaces. This property is called *adsorption.* The use of charcoal in sugar refining to adsorb coloring matter is based on this property.

Colloidal particles usually carry an electric charge due to the adsorption of ions from the dispersion medium and to the ionization of the colloidal particles themselves. Colloids selectively adsorb ions from a solution; therefore all particles of a given colloid carry like charges.

The kind of charge carried by the colloidal particles may be determined by *cataphoresis* or *electrophoresis*. The colloidal solution to be tested is placed in a U-tube fitted with electrodes. As an electric current is passed through, the colloid will migrate to the anode if it is negatively charged and to the cathode if it is positively charged.

Colloids are retained by a semipermeable membrane, which allows crystalloids to pass through. Therefore colloids can be separated from crystalloids by means of a semipermeable membrane, a process called *dialysis*. The process of absorption of digested food material through the mucous membrane of the intestinal wall is an example of dialysis. This is also the principle upon which the artificial kidney operates. As the patient's blood flows through the artificial kidney, its waste products are dialyzed out and clean blood flows back to the patient.

Lyophilic colloids can act as protective colloids for lyophobic colloids because they prevent them from being precipitated out by oppositely charged ions.

# 10. ELECTROLYTES AND IONIZATION

Chemical compounds can be divided into two categories, electrolytes and nonelectrolytes, depending on whether or not they undergo ionization in solution.

**Electrolytes.** Electrolytes are those substances whose water solutions conduct electric current as a result of ionization or dissociation. Acids, bases, and salts are electrolytes.

*Strong electrolytes* include strong acids such as hydrochloric, nitric, and sulfuric; strong bases such as sodium hydroxide and potassium hydroxide; and most salts whose water solutions have high conductivity and a high degree of dissociation. The equation below shows how a solution of hydrochloric acid dissociates into positively charged hydrogen ions and negatively charged chloride ions.

$$HCl \rightarrow H^+ + Cl^-$$

*Weak electrolytes* include weak acids such as acetic, boric, and carbonic, and weak bases such as ammonium hydroxide, whose water solutions have low conductivity and a low degree of dissociation. The following equation shows the dissociation of a solution of acetic acid into positively charged hydrogen ions and negatively charged acetate ions.

$$HC_2H_3O_2 \rightleftharpoons H^+ + C_2H_3O_2^-$$

The double arrow means that the ionization of a weak electrolyte is a reversible reaction. The short arrow indicates that, at equilibrium, there are less ions than nonionized molecules.

The percentage of ionization of some typical electrolytes in dilute solution is shown below.

| ACIDS | | PERCENTAGE OF IONIZATION |
|---|---|---|
| Hydrochloric acid | $HCl$ | 92.0 |
| Nitric acid | $HNO_3$ | 92.0 |
| Sulfuric acid | $H_2SO_4$ | 61.0 |
| Acetic acid | $HC_2H_3O_2$ | 1.3 |
| Carbonic acid | $H_2CO_3$ | 0.17 |
| Boric acid | $H_3BO_3$ | 0.01 |

BASES

| Potassium hydroxide | KOH | 91.0 |
| Sodium hydroxide | NaOH | 91.0 |
| Ammonium hydroxide | $NH_4OH$ | 1.3 |

MOST SALTS          70–100

*Nonelectrolytes* are those substances whose water solutions do not conduct electric current. They include most of the organic compounds (e.g., alcohols, aldehydes, ketones, and sugar), which do not ionize in water solution. They are compounds in which the atoms are held together by covalent bonds, whereas electrolytes have electrovalent bonds.

**Arrhenius Theory of Ionization.** This theory was proposed by the Swedish chemist Arrhenius in 1887. The main assumptions of this theory are:

1. When an electrolyte is dissolved in water some of its molecules break up into positively charged particles and negatively charged particles called *ions*.

2. The positively charged ions are called *cations* because they are attracted to the cathode. The negatively charged ions are called *anions* because they are attracted to the anode.

3. The sum of the positive charges is equal to the sum of the negative charges, so that the solution as a whole is electrically neutral.

4. The degree of ionization or dissociation is dependent upon the concentration of the solution; the more dilute the solution, the more completely the electrolyte dissociates into ions. This is because in dilute solutions the ions formed meet less frequently and are therefore less likely to unite into molecules.

**Debye-Hückel Theory of Strong Electrolytes.** Since the particles that make up salt crystals are ions rather than molecules, salts are 100 percent ionized in solution. However, they appear to be only partially ionized. This apparent partial ionization is due to *interionic attraction* of oppositely charged ions which restrict each other's movement in solution.

**Hydrolysis.** Certain salts will react with water by double decomposition to form acid or alkaline solutions. The reaction is called *hydrolysis.*

Salts of weak bases and strong acids yield acid solutions, because the weak base that is formed yields very few hydroxyl ions ($OH^-$), while the strong acid that is formed yields many hydrogen ions ($H^+$). The hydrolysis of ammonium chloride is shown below.

$$NH_4^+Cl^- + HOH \rightarrow NH_4OH + H^+Cl^-$$

| Ammonium chloride | | Weak base | Strong acid |

Salts of strong bases and weak acids yield alkaline solutions, because the weak acid that is formed yields very few hydrogen ions ($H^+$), while the strong base that is formed yields many hydroxyl ions ($OH^-$). The following equation shows the hydrolysis of sodium acetate.

$$Na^1C_2H_3O_2^- \mid HOH \rightarrow HC_2H_3O_2 + Na^+OH^-$$

| Sodium acetate | | Weak acid | Strong base |

**Hydrogen Ion Concentration.** Water dissociates very slightly, only to the extent of 1 mole in every 10,000,000 liters. The concentration of hydrogen ions is therefore $1/10,000,000$ or $1/10^7$ or $10^{-7}$ mole per liter, and the concentration of hydroxyl ions is also $1/10,000,000$ or $1/10^7$ or $10^{-7}$ mole per liter.

The equation below shows the ionization of water.

$$H_2O \rightleftharpoons H^+ + OH^-$$

It has been found that whether the solution is acid, alkaline, or neutral, the hydrogen ion concentration multiplied by the hydroxyl ion concentration always equals $10^{-14}$, the ionic product of water.

Since the product of the two ion concentrations always equals $10^{-14}$, any increase in the concentration of the hydrogen ions by the addition of acid, say from $10^{-7}$ to $10^{-4}$, will result in a corresponding decrease in the hydroxyl ion concentration from $10^{-7}$ to $10^{-10}$.

Conversely, if the concentration of the hydroxyl ions is increased, say, to $10^{-2}$ by the addition of base, the concentration of the hydrogen ions will be decreased to $10^{-12}$.

THE pH SCALE. Hydrogen ion concentration is more easily expressed by leaving out $10^-$ and using only the numerical value of the negative power and calling it pH. Mathematically, $pH = -\log H^+$ concentration. For example,

$$10^{-7} = pH\ 7\ \ neutral$$
$$10^{-4} = pH\ 4\ \ acid$$
$$10^{-10} = pH\ 10\ \ alkaline$$

## TABLE 10.1.   THE pH RANGE OF INDICATORS AND BODY FLUIDS

| H⁺ Concentration | $10^{-1}$ | $10^{-2}$ | $10^{-3}$ | $10^{-4}$ | $10^{-5}$ | $10^{-6}$ | $10^{-7}$ | $10^{-8}$ | $10^{-9}$ | $10^{-10}$ | $10^{-11}$ | $10^{-12}$ | $10^{-13}$ |
|---|---|---|---|---|---|---|---|---|---|---|---|---|---|
| pH | 1 | 2 | 3 | 4 | 5 | 6 | 7 | 8 | 9 | 10 | 11 | 12 | 13 |
| Reaction | ACID | | | | | | NEU-TRAL | ALKALINE | | | | | |
| | | | | | | | INDICATORS | | | | | | |
| Methyl Orange | | | red ⟵⟶ yellow | | | | | | | | | | |
| Litmus | | | | | red ⟵⟶ blue | | | | | | | | |
| Phenolphthalein | | | | | | | colorless ⟵⟶ red | | | | | | |
| | | | | | | | BODY FLUIDS | | | | | | |
| Gastric juice | ⟵⟶ | | | | | | | | | | | | |
| Sweat | | | | ↔ | | | | | | | | | |
| Urine | | | | | ⟵⟶ | | | | | | | | |
| Saliva | | | | | | ⟵⟶ | | | | | | | |
| Blood | | | | | | | ↔ | | | | | | |
| Intestinal juice | | | | | ⟵⟶ | | | | | | | | |
| Bile | | | | | | | ⟵⟶ | | | | | | |
| Pancreatic juice | | | | | | | ↔ | | | | | | |

Thus, if the pH is 7.0, the solution is neutral; above 7.0 it is alkaline and below 7.0 it is acid. See Table 10.1.

DETERMINATION OF pH.   The pH of a solution can be determined by one of two methods, the colorimetric method and the electrometric method.

*Colorimetric Method.*   The colorimetric method of measuring pH uses a set of indicators which show definite color shades for each pH value.   By adding a suitable indicator to the solution and matching its color with that of one of the standard pH solutions containing the same indicator, the pH of the solution can be estimated to the nearest 0.1 pH unit.   A list of commonly used indicators is given in Table 10.2.

A more rough but simpler method of determining the pH of a solution is to place a drop of the solution on a strip of indicator paper and match it with a color standard furnished with the test paper.

*Electrometric Method.*   In the electrometric, or potentiometric, method, an instrument called the pH meter is employed.   The two electrodes of the meter are dipped into the solution to be tested and the pH is then read directly from the dial of the meter.

The electrometric method is more accurate than the colori-

## TABLE 10.2.    ACID-BASE INDICATORS

| Indicator | pH Range | Color | |
|---|---|---|---|
| | | At Lower pH | At Higher pH |
| Methyl violet | 0.2–3.2 | Yellow | Violet |
| Thymol blue | 1.2–2.8 | Red | Yellow |
| Methyl orange | 3.1–4.4 | Red | Yellow |
| Bromphenol blue | 3.0–4.6 | Yellow | Blue |
| Bromcresol green | 4.0–5.6 | Yellow | Blue |
| Methyl red | 4.3–6.1 | Red | Yellow |
| Chlorphenol red | 4.8–6.4 | Yellow | Red |
| Bromcresol purple | 5.2–6.8 | Yellow | Purple |
| Bromthymol blue | 6.0–7.6 | Yellow | Blue |
| Phenol red | 6.8–8.4 | Yellow | Red |
| Cresol red | 7.2–8.8 | Yellow | Red |
| Thymol blue | 8.0 9.6 | Yellow | Blue |
| Phenolphthalein | 8.3–10.0 | Colorless | Red |
| Alizarin yellow | 10.1–12.1 | Yellow | Violet |
| Nitramine | 11.0–13.0 | Colorless | Brown |

metric method.  It is the only method that can be used when the solution to be tested is highly turbid or deeply colored.

pH meters are used extensively in industry and in biological and medical research laboratories.

**Buffers.**  Buffers are substances which in solution will resist appreciable changes in pH upon addition of either acids or bases. They consist of a weak acid and its salt or a weak base and its salt, such as the buffer pair $H_2CO_3/NaHCO_3$ (carbonic acid and sodium bicarbonate) found in the blood.  This buffer system neutralizes acid by the reaction of its salt with the acid:

$$Na^+HCO_3^- + H^+Cl^- \rightarrow H_2CO_3 + Na^+Cl^-$$

It neutralizes base by the reaction of its weak acid with the base:

$$H_2CO_3 + Na^+OH^- \rightarrow H_2O + Na^+HCO_3^-$$

The $H_2CO_3/NaHCO_3$ buffer system helps maintain the pH of the blood at about 7.4.  Buffer systems are found in all body fluids.

# 11. ACIDS, BASES, AND SALTS

The majority of inorganic compounds can be classified as acids, bases, or salts. Acids and bases each contain a common element or radical and possess a distinct set of properties.

## ACIDS

An acid is a chemical compound that yields the hydrogen ion ($H^+$) in solution. According to modern concepts, first proposed by Brönsted and Lowry, the hydrogen ion, which is actually a proton, does not exist in water solutions. It combines with one molecule of water to form the hydronium ion ($H_3O^+$). However, for convenience in discussion, we shall refer to the hydrogen ion instead of the more complicated hydronium ion.

The Brönsted-Lowry theory defines an acid as a proton donor, that is, any substance which can release a proton (hydrogen ion). Thus, by this definition, HCl, $H_2O$, and $NH_4^+$ are acids:

$$HCl \rightarrow H^+ + Cl^-$$
$$H_2O \rightarrow H^+ + OH^-$$
$$NH_4^+ \rightarrow H^+ + NH_3$$

**Classification of Acids.** Acids are classified according to the maximum number of hydrogen ions or protons one molecule can yield in solution. Acids which liberate one hydrogen ion are called *monobasic* or monoprotic acids (e.g., HCl, $HNO_3$); those which liberate two hydrogen ions are called *dibasic* or diprotic acids (e.g., $H_2SO_4$, $H_2CO_3$); those which yield three hydrogen ions are *tribasic* or triprotic acids (e.g., $H_3PO_4$, $H_3BO_3$).

**Properties of Acids.** Acids may be solids, liquids, or gases. They have a sour taste. They turn blue litmus red.

REACTION WITH METALS. Acids react with active or moderately active metals like sodium, potassium, calcium, and zinc to liberate hydrogen. The following equation shows that zinc replaces the hydrogen of sulfuric acid to form zinc sulfate and hydrogen gas.

$$Zn + H_2SO_4 \rightarrow ZnSO_4 + H_2 \uparrow$$

To show which metals will, and which metals will not react with acids to liberate hydrogen and how their reactivity compares with each other, the common metals have been arranged in what is known as the *activity series*, shown below.

K
Na
Ca
Mg
Al
Zn
Fe
Ni
Sn
Pb
H
Cu
Hg
Ag
Pt
Au

1. Metals near the top of the series are the most active, capable of liberating hydrogen even from water.

2. All metals above hydrogen will liberate hydrogen from acids; all metals below hydrogen will not.

3. Any metal will replace any metal below it from its salt in solution.

REACTION WITH OXIDES.  Acids react with oxides to form salts and water.

$$2HCl + MgO \rightarrow MgCl_2 + H_2O$$

REACTION WITH BASES.  Acids react with bases to form salts and water. This type of reaction is called *neutralization*.

$$HCl + NaOH \rightarrow NaCl + H_2O$$

ACID-BASE TITRATIONS.  By knowing the concentration of either the acid or the base, the concentration of the other can be determined by titrating the two solutions. *Titration* is the process of adding an acid solution to a basic solution (or of a basic solution to an acid solution) to determine the point at which complete neutralization occurs. To perform a titration, a known volume of the base of unknown concentration is placed in a flask. A few drops of an indicator, such as phenolphthalein, which has one color in an acid solution and another color in a basic solution, is introduced. A standard acid solution of known concentration is then added from a buret until the neutralization of the base is complete, at which point, called the *end point*, the indicator changes color. Knowing the volume of the standard acid solution used, the concentration of the base can be calculated. In the same manner, the concentration of an acid can be determined by titrating it with a standard base.

*Problem:*   If it takes 30 ml. of a 0.1N solution of sodium hydroxide to neutralize 25 ml. of a solution of hydrochloric acid of unknown concentration, what is the concentration of the hydrochloric acid?

Since the smaller the normality, the greater the volume required for neutralization, the volumes of the solutions are inversely proportional to their concentrations. Hence:

$$\frac{30 \text{ ml.}}{25 \text{ ml.}} = \frac{x \text{ N}}{0.1\text{N}}$$

$X$ = 0.12N (normality of the HCl)

REACTION WITH CARBONATES.   Acids react with carbonates and bicarbonates to form carbon dioxide, water, and salt.

$$H_2SO_4 + Na_2CO_3 \rightarrow CO_2\uparrow + H_2O + Na_2SO_4$$
Sodium
carbonate

$$HCl + NaHCO_3 \rightarrow CO_2\uparrow + H_2O + NaCl$$
Sodium
bicarbonate

**Naming of Acids.**   All acids contain hydrogen; most acids contain oxygen as well.  Acids composed of hydrogen and one other element are called *binary* or hydro acids.  Such acids are given the prefix *hydro* and the suffix *ic.*

| | | | | |
|---|---|---|---|---|
| HCl | *Hydro* | chlor | *ic* | acid |
| HBr | *Hydro* | brom | *ic* | acid |

Acids that contain oxygen and another element in addition to hydrogen are called *ternary* or oxyacids.   The most common ternary acids end in *ic.*

| | | | |
|---|---|---|---|
| $H_2SO_4$ | Sulfur | *ic* | acid |
| $HClO_3$ | Chlor | *ic* | acid |

Some ternary acids have the same elements but vary in oxygen content.  They are named as follows:

Ternary acids containing less oxygen end in *ous.*

| | | | |
|---|---|---|---|
| $H_2SO_3$ | Sulfur | *ous* | acid |
| $HClO_2$ | Chlor | *ous* | acid |

Ternary acids containing the least amount of oxygen carry the prefix *hypo* and the suffix *ous.*

$$HClO \quad \textit{Hypo} \quad \text{chlor} \quad \textit{ous} \quad \text{acid}$$

Ternary acids containing more oxygen than the *ic* acids carry the prefix *per* and the suffix *ic*.

$$HClO_4 \quad \textit{Per} \quad \text{chlor} \quad \textit{ic} \quad \text{acid}$$

## BASES

A base is a chemical compound that yields the hydroxide ion ($OH^-$) in solution. According to the Brönsted-Lowry theory, a base is a proton acceptor, that is, any substance which can accept a proton (hydrogen ion). By this definition, $Cl^-$, $OH^-$, and $NH_3$ are bases:

$$Cl^- + H^+ \rightarrow HCl$$
$$OH^- + H^+ \rightarrow H_2O$$
$$NH_3 + H^+ \rightarrow NH_4^+$$

**Properties of Bases.**   Most bases are solids. Solutions of bases have a bitter taste and a slippery feeling. They turn red litmus blue.

Bases react with acids to form salts and water.

$$2KOH + H_2SO_4 \rightarrow K_2SO_4 + 2H_2O$$

**Naming of Bases.**   Bases are named by starting with the name of the metal (or the ammonium radical) of which it is composed and ending with the word *hydroxide*.

NaOH      Sodium *hydroxide*
$NH_4OH$   Ammonium *hydroxide*

When metals having two valences combine with the hydroxyl radical to form bases, the suffix *ous* is applied to the base in which the lower valence appears and the suffix *ic* to the base showing the higher valence.

$Fe(OH)_2$   Ferr   *ous*   hydroxide
$Fe(OH)_3$   Ferr   *ic*   hydroxide

## SALTS

A salt is a chemical compound that is made up of a positive ion, other than hydrogen, and a negative ion, other than the hydroxide ion. It is formed by the reaction of an acid and a base;

its positive ion is derived from the base and its negative ion from the acid. Besides being electrolytes, salt solutions have no common properties because there is no ion common to all salts.

**Classification of Salts.** Salts are called *normal* when they contain neither replaceable hydrogen nor the hydroxide ion.

$$\text{Sodium carbonate} \quad Na_2CO_3$$

*Acid salts* contain replaceable hydrogen.

Sodium bicarbonate
(sodium acid carbonate)    $NaHCO_3$
Monosodium acid phosphate    $NaH_2PO_4$

*Basic salts* contain replaceable hydroxide group.

Basic bismuth nitrate    $Bi(OH)_2NO_3$
Basic lead carbonate    $Pb(OH)_2 \cdot 2PbCO_3$
(white lead)

*Mixed salts* contain more than one positive ion (or metal).

Potassium sodium tartrate    $KNaC_4H_4O_6$
(Rochelle salt)

*Double salts* contain two distinct cations or anions other than hydrogen or hydroxyl groups.

Calcium magnesium carbonate (dolomite)    $CaMg(CO_3)_2$
Calcium oxychloride (chloride of lime)    $Ca(OCl)Cl$

Double salts that contain equi-molar amounts of sulfates of monovalent and trivalent metal ions are known as *alum*

Potassium alum    $K_2SO_4 \cdot Al_2(SO_4)_3 \cdot 24H_2O$
Ammonium alum    $(NH_4)_2SO_4 \cdot Al_2(SO_4)_3 \cdot 24H_2O$

**Naming of Salts.** The first part of the name of a salt is the name of the metal of the base from which it was formed; the second part is the name of the nonmetallic element or radical of the acid from which it was derived.

*Binary salts* are named from the corresponding acids by changing *ic* to *ide*.

HCl    Hydrochlor *ic* acid
NaCl    Sodium chlor *ide*

In the above example, the *sodium* comes from the name of the base, sodium hydroxide, from which this salt was formed.

*Ternary salts* are named from the corresponding acids as follows: If the name of the acid ends in *ic*, the name of the salt ends in *ate*.

$$H_2SO_4 \quad \text{sulfur} \quad ic \quad \text{acid}$$
$$CaSO_4 \quad \text{calcium} \quad \text{sulf} \quad ate$$

If the name of the acid ends in *ous*, the name of the salt ends in *ite*.

$$H_2SO_3 \quad \text{sulfur} \quad ous \quad \text{acid}$$
$$CaSO_3 \quad \text{calcium} \quad \text{sulf} \quad ite$$

In like manner a salt derived from a *hypo ous* acid becomes *hypo ite*

$$HClO \quad hypo \quad \text{chlor} \quad ous \quad \text{acid}$$
$$NaClO \quad \text{sodium} \quad hypo \quad \text{chlor} \quad ite$$

and a salt derived from a *per ic* acid becomes *per ate*.

$$HClO_4 \quad per \quad \text{chlor} \quad ic \quad \text{acid}$$
$$KClO_4 \quad \text{potassium} \quad per \quad \text{chlor} \quad ate$$

**Uses of Salts.**   The following is a list of some of the common salts that find important uses in the home and in the practice of medicine.

| Salt | Formula | Uses |
|---|---|---|
| Aluminum sulfate | $Al_2(SO_4)_3$ | Deodorant, water purification |
| Ammonium chloride | $NH_4Cl$ | Diuretic, expectorant |
| Barium sulfate | $BaSO_4$ | X-ray studies |
| Calcium chloride | $CaCl_2$ | Intravenous administration |
| Calcium sulfate (plaster of Paris) | $(CaSO_4)_2 \cdot H_2O$ | Casts, molds |
| Magnesium carbonate | $MgCO_3$ | Antacid, cosmetic |
| Magnesium sulfate (Epsom salt) | $MgSO_4 \cdot 7H_2O$ | Purgative |
| Mercuric chloride (corrosive sublimate) | $HgCl_2$ | Antiseptic, disinfectant |
| Potassium chloride | $KCl$ | Intravenous administration |
| Silver nitrate (lunar caustic) | $AgNO_3$ | Antiseptic |
| Sodium bicarbonate (baking soda) | $NaHCO_3$ | Baking, antacid |
| Sodium carbonate (washing soda) | $Na_2CO_3$ | Water softener, soap powder |
| Sodium chloride (table salt) | $NaCl$ | Seasoning, physiological saline |
| Sodium iodide | $NaI$ | Iodized table salt |
| Sodium hypochlorite | $NaClO$ | Bleaching solution |

# 12. ORGANIC CHEMISTRY

Organic chemistry is the study of the compounds of carbon. For centuries it was believed that all organic compounds were produced in the cells of plants and animals through the aid of a "vital force" within living organisms. However, in 1828 Friedrich Wöhler, a German chemist, succeeded in producing urea, an organic substance formed in the body, by heating the inorganic substance ammonium cyanate.

$$NH_4OCN \longrightarrow H_2NCONH_2$$

Ammonium        Urea
cyanate

Since then, thousands of organic compounds have been prepared from inorganic sources.

Organic chemistry is involved in practically all phases of modern life, since it deals with food, vitamins, hormones, enzymes, disinfectants, anesthetics, drugs, fuel, flavors, perfumes, dyes, explosives, detergents, plastics, rubber, and natural and synthetic fibers.

**Comparison of Organic and Inorganic Compounds.**  There are many more organic compounds than inorganic compounds. Whereas any of the elements may combine to form inorganic substances, organic compounds always contain carbon, most of them hydrogen, and many of them oxygen. Other elements found in organic compounds are, in relative order of occurrence, nitrogen, phosphorus, sulfur, chlorine, bromine, and iodine.  The molecules of organic compounds are usually more complex and have a higher molecular weight than those of inorganic compounds.

Another important difference between organic and inorganic compounds is the type of bonding.  Most inorganic compounds are electrovalent and ionize in solution, while most organic compounds are covalent and nonelectrolytes.  Thus the reactions of organic compounds are slow compared to those of inorganic compounds. The type of bonding also accounts for the characteristic differences in physical properties between organic and inorganic compounds. See Table 12.1.

TABLE 12.1.   COMPARISON OF ORGANIC AND INORGANIC
COMPOUNDS

| | Organic Compounds | Inorganic Compounds |
|---|---|---|
| Number of known compounds | Over 500,000 | About 30,000 |
| Structure | Often complex with high molecular weights | Generally simple with comparatively low molecular weights |
| Type of bonding | Mostly covalent | Mostly ionic or electrovalent |
| Melting point and boiling point | Generally low | Generally high |
| Solubility | Soluble in organic solvents | Soluble in water |
| Conductivity of water solutions | Mostly nonconductors | Mostly conductors |
| Speed of reactions | Slow, because reactions are between molecules | Fast, because reactions are between ions |
| Flammability | Usually flammable | Usually nonflammable |

**Properties of Carbon.**   Elements with less than four electrons in the outer shell have the tendency to lose those electrons to form positively charged ions.   Elements with more than four electrons in the outer shell have the tendency to gain more electrons to complete the octet and form negatively charged ions. The carbon atom, which has four electrons in its outer shell, has the tendency to share its electrons with other atoms in forming compounds.   For example, carbon will share its electrons with four hydrogen atoms, forming methane.

$$\cdot \overset{\cdot}{\underset{\cdot}{C}} \cdot \; + \; 4H \cdot \; \rightarrow \; H \overset{H}{\underset{H}{:C:}} H$$

Methane

The above electron dot formula for methane may be simplified by using a single bond (straight line) to represent a pair of shared electrons.   Thus the structural formula for methane is

$$H - \overset{\displaystyle H}{\underset{\displaystyle H}{\overset{|}{\underset{|}{C}}}} - H$$

In space the four valence bonds in methane do not extend from the carbon atom in one plane as indicated above, but extend to each corner of a three-dimensional tetrahedron as shown below.

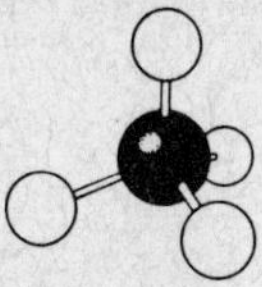

Carbon is the only element that can form bonds with itself. Thus the carbon atom will share its electrons with other carbon atoms to form chains and rings. This unique capacity of carbon accounts for the large number of possible organic compounds. Some examples are shown below.

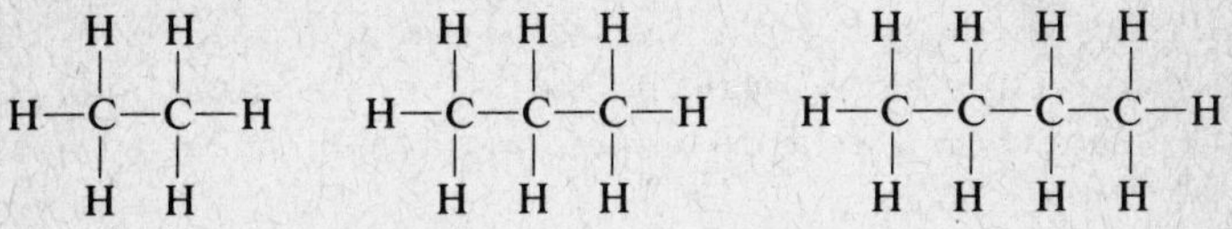

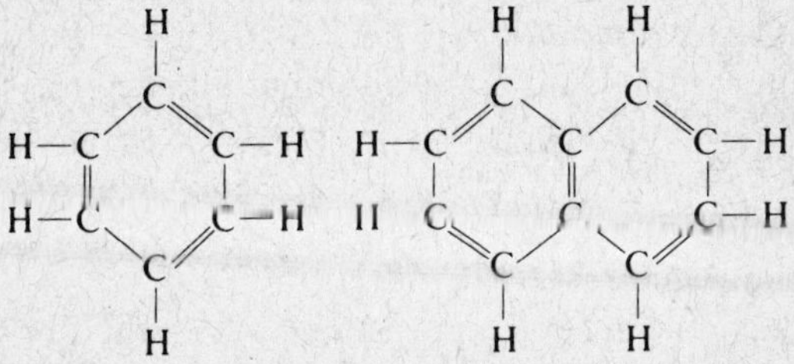

Another factor contributing to the large number of organic compounds is the phenomenon called *isomerism*. Isomers are two or more compounds of different physical and chemical properties having the same molecular formula but different structural formulas. For example, ethyl alcohol and dimethyl ether are isomers.

Molecular Formula  $C_2H_6O$

Structural Formulas

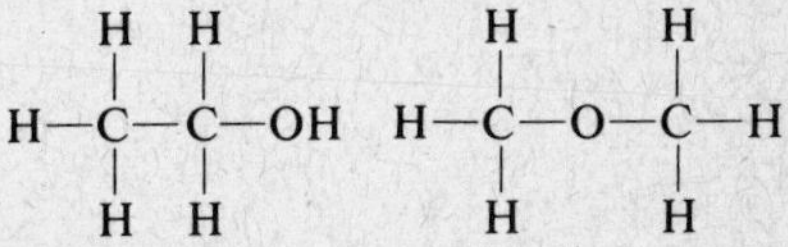

Ethyl alcohol            Dimethyl ether

# 13. ALIPHATIC HYDROCARBONS

Hydrocarbons are compounds containing only hydrogen and carbon. The aliphatic hydrocarbons are in the form of chains which may be straight or branched.

## ALKANES

Alkanes are those hydrocarbons in which the carbon atoms are joined to each other by single bonds. They are called *saturated hydrocarbons* because the carbon atoms are completely saturated with hydrogen atoms.

They are also called *paraffin hydrocarbons* because of their lack of affinity. The word *paraffin* is derived from the Latin, *parvum affinis*, meaning little affinity. In addition, alkanes are sometimes referred to as the *methane hydrocarbons* because methane is the first or parent compound of the *homologous series*, a series in which each succeeding compound contains $CH_2$ more than the preceding compound. See Table 13.1.

Alkanes have the general formula $C_n H_{2n+2}$, where $n$ stands for the number of carbon atoms.

TABLE 13.1.   NORMAL PARAFFIN HYDROCARBONS*

| Name | Molecular Formula | Melting Point | Boiling Point | Calculated No. of Isomers |
|------|------|------|------|------|
| Methane | $CH_4$ | $-182.6°C.$ | $-161.7°C.$ | 1 |
| Ethane | $C_2H_6$ | $-172.0$ | $-88.6$ | 1 |
| Propane | $C_3H_8$ | $-187.1$ | $-42.2$ | 1 |
| Butane | $C_4H_{10}$ | $-135.0$ | $-0.5$ | 2 |
| Pentane | $C_5H_{12}$ | $-129.7$ | $36.1$ | 3 |
| Hexane | $C_6H_{14}$ | $-94.0$ | $68.7$ | 5 |
| Heptane | $C_7H_{16}$ | $-90.5$ | $98.4$ | 9 |
| Octane | $C_8H_{18}$ | $-56.8$ | $125.6$ | 18 |
| Nonane | $C_9H_{20}$ | $-53.7$ | $150.7$ | 35 |
| Decane | $C_{10}H_{22}$ | $-29.7$ | $174.0$ | 75 |

*For the structures of some simple isomers, see Table 13.2.

## TABLE 13.2. STRUCTURAL FORMULAS OF SIMPLE METHANE HYDROCARBONS

| Hydrocarbon | Molecular Formula | Structural Formula |
|---|---|---|
| Methane | $CH_4$ | 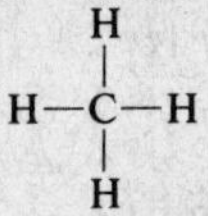 $H-\underset{\underset{H}{\displaystyle\vert}}{\overset{\overset{H}{\displaystyle\vert}}{C}}-H$ |
| Ethane | $C_2H_6$ | 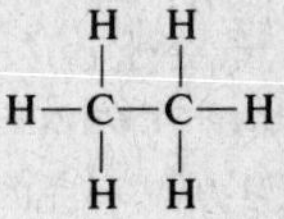 $H-\overset{H}{\underset{H}{C}}-\overset{H}{\underset{H}{C}}-H$ |
| Propane | $C_3H_8$ | 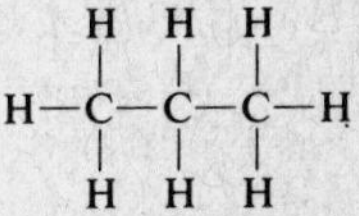 $H-\overset{H}{\underset{H}{C}}-\overset{H}{\underset{H}{C}}-\overset{H}{\underset{H}{C}}-H$ |
| n-Butane | $C_4H_{10}$ | 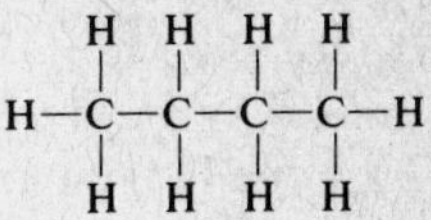 $H-\overset{H}{\underset{H}{C}}-\overset{H}{\underset{H}{C}}-\overset{H}{\underset{H}{C}}-\overset{H}{\underset{H}{C}}-H$ |
| Isobutane (2-methylpropane) | $C_4H_{10}$ | 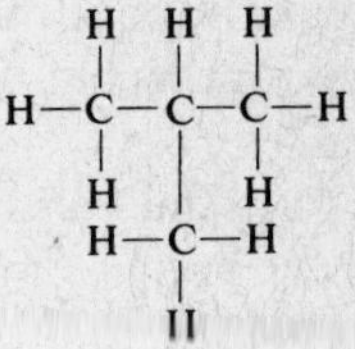 $H-\overset{H}{\underset{H}{C}}-\overset{H}{\underset{\underset{H-C-H}{H}}{C}}-\overset{H}{\underset{H}{C}}-H$ |
| n-Pentane | $C_5H_{12}$ | 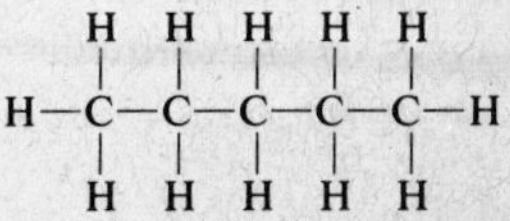 $H-\overset{H}{\underset{H}{C}}-\overset{H}{\underset{H}{C}}-\overset{H}{\underset{H}{C}}-\overset{H}{\underset{H}{C}}-\overset{H}{\underset{H}{C}}-H$ |
| Isopentane (2-methylbutane) | $C_5H_{12}$ | 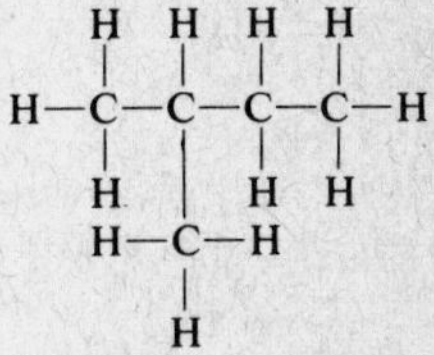 $H-\overset{H}{\underset{H}{C}}-\overset{H}{\underset{\underset{H-C-H}{H}}{C}}-\overset{H}{\underset{H}{C}}-\overset{H}{\underset{H}{C}}-H$ |
| Neopentane (2,2-dimethylpropane) | $C_5H_{12}$ | 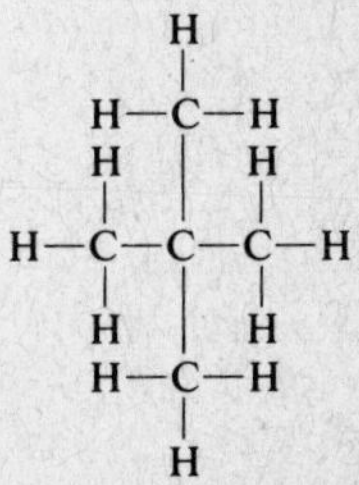 $H-\overset{H}{\underset{H}{C}}-\overset{H-\overset{H}{\underset{H}{C}}-H}{\underset{H-\underset{H}{\overset{H}{C}}-H}{C}}-\overset{H}{\underset{H}{C}}-H$ |

**Naming of Alkanes.** All the alkanes end in *ane* and all the alkyl radicals (containing one less hydrogen than the alkanes) end in *yl*.

| Alkane Hydrocarbons | | Alkyl Radicals | |
|---|---|---|---|
| $CH_4$ | methane | $CH_3-$ | methyl |
| $C_2H_6$ | ethane | $C_2H_5-$ | ethyl |
| $C_3H_8$ | propane | $C_3H_7-$ | propyl |
| $C_4H_{10}$ | butane | $C_4H_9-$ | butyl |

The next members of the series are, in order, pentane, hexane, heptane, octane, nonane, decane, etc.

The straight-chain hydrocarbons are called normal or *n-*. Thus $CH_3-CH_2-CH_2-CH_3$ is called normal butane, or *n*-butane. *n*-Butane has a branched-chain isomer called isobutane, and pentane has two isomers, known as isopentane and neopentane. When there is more than one isomer, the compounds are best named according to the Geneva or I.U.C. (International Union of Chemistry) system as follows:

1. Find the longest carbon chain in the structural formula of the molecule. Count the carbons in that chain. Give the compound the root name of the normal alkane having that number of carbons.

2. Number the carbons in the chain, beginning at the end closest to the branches or side-chains.

3. Precede the root name by the radical names of the side-chains, which in turn are preceded by the number of the carbon atom in the main chain to which the radical is attached.

$$\overset{1}{C}H_3-\overset{2}{C}H-\overset{3}{C}H_2-\overset{4}{C}H_3$$
$$\quad\quad|$$
$$\quad\quad CH_3$$

2-methylbutane

$$\overset{1}{C}H_3-\overset{2}{C}H-\overset{3}{C}H_2-\overset{4}{C}H-\overset{5}{C}H_3$$
$$\quad\quad|\quad\quad\quad\quad|$$
$$\quad\quad CH_3\quad\quad\quad CH_3$$

2,4-dimethylpentane

$$\quad\quad\quad\quad CH_3$$
$$\quad\quad\quad\quad|$$
$$CH_3-CH_2-C-CH_2-CH_2-CH_3$$
$$\quad\quad\quad\quad|$$
$$\quad\quad\quad\quad CH_3$$

3,3-dimethylhexane

$$CH_3-CH-CH_2-CH_2-CH-CH_2-CH_3$$
$$\quad\quad|\quad\quad\quad\quad\quad\quad|$$
$$\quad\quad CH_3\quad\quad\quad\quad\quad CH_2$$
$$\quad\quad\quad\quad\quad\quad\quad\quad|$$
$$\quad\quad\quad\quad\quad\quad\quad\quad CH_3$$

2-methyl-5-ethylheptane

**Physical Properties of Alkanes.** At ordinary temperatures the alkanes $C_1$ to $C_4$ are gases, $C_5$ to $C_{15}$ are liquids, and $C_{16}$ and above are solids. The alkanes are insoluble in water but soluble in organic solvents such as ether. Boiling point and specific gravity increase with the number of carbon atoms.

**Chemical Properties of Alkanes.** Alkanes are the least reactive group of organic compounds. They are resistant to attack by common laboratory reagents such as sulfuric acid, sodium hydroxide, and potassium dichromate or permanganate.

In the presence of light, the alkanes react with chlorine or bromine by substitution, in which one or more hydrogens of the hydrocarbon may be substituted by the halogen.

$$CH_4 + Cl_2 \rightarrow HCl + CH_3Cl$$
Methane         Methyl chloride

$$CH_3Cl + Cl_2 \rightarrow HCl + CH_2Cl_2$$
Methylene chloride

$$CH_2Cl_2 + Cl_2 \rightarrow HCl + CHCl_3$$
Chloroform

$$CHCl_3 + Cl_2 \rightarrow HCl + CCl_4$$
Carbon
tetrachloride

The alkanes burn in air with the formation of carbon dioxide and water and the production of energy, which may be used in heating, cooking, and in running the automobile or airplane.

$$C_7H_{16} + 11O_2 \rightarrow 7CO_2 + 8H_2O + energy$$
Heptane

**Sources and Uses of Hydrocarbons.** *Natural gas* consists largely of methane and smaller amounts of other low molecular weight hydrocarbons up to $C_8$. The chief use of natural gas is as a fuel. *Petroleum* yields, by fractional distillation, higher molecular weight hydrocarbons that are of utmost importance to the home and transportation. See Table 13.3.

Besides the fraction obtained from the first distillation, referred to as straight-run gasoline, large amounts of gasoline are obtained by *cracking*, which is the breaking down of larger molecules into smaller ones. In *thermal cracking* the vapors of high-boiling fractions are passed through a hot (500–700°C.) metal tube. In *catalytic cracking* the oil vapors are heated in the presence of silica and alumina.

TABLE 13.3.   PETROLEUM FRACTIONS

| Fraction | Molecular Size | Boiling Range (°C.) | Uses |
|---|---|---|---|
| Gases | $C_1$—$C_4$ | Below 20 | Fuel |
| Petroleum ether and naphtha | $C_5$—$C_7$ | 20–100 | Solvents |
| Gasoline | $C_6$—$C_{12}$ | 40–200 | Motor fuel |
| Kerosene | $C_{12}$—$C_{15}$ | 175–275 | Illuminent, fuel |
| Fuel oil | $C_{15}$ and higher | 250–400 | Fuel |
| Lubricating oils and residue | $C_{19}$ and up | Above 300 | Lubricants, asphalt, paraffin |

## ALKENES

The alkenes contain two hydrogen atoms less than the corresponding alkanes and are therefore unsaturated. Thus the alkenes contain two carbon atoms joined by a double bond. They are called *olefins* because ethylene, the first member of the alkene series, unites with chlorine or bromine to give an oily product (L. *oleum*, oil, and *fiant*, making). They have the general formula $C_nH_{2n}$ and the ending *ylene* or *ene*.

$$C_2H_4 \qquad \underset{\text{Ethylene or ethene}}{H-\overset{\overset{\displaystyle H}{|}}{C}=\overset{\overset{\displaystyle H}{|}}{C}-H}$$

$$C_3H_6 \qquad \underset{\text{Propylene or propene}}{H-\underset{\underset{\displaystyle H}{|}}{\overset{\overset{\displaystyle H}{|}}{C}}-\overset{\overset{\displaystyle H}{|}}{C}=\overset{\overset{\displaystyle H}{|}}{C}-H}$$

Alkenes with two double bonds are called *dienes*.

$$C_4H_6 \qquad H-\overset{\overset{\displaystyle H}{|}}{C}=\overset{\overset{\displaystyle H}{|}}{C}-\overset{\overset{\displaystyle H}{|}}{C}=\overset{\overset{\displaystyle H}{|}}{C}-H$$

Butadiene

The alkenes may be prepared by the dehydration of alcohols in the presence of concentrated sulfuric acid.

$$H-\overset{\overset{\displaystyle H}{|}}{\underset{\underset{\displaystyle H}{|}}{C}}-\overset{\overset{\displaystyle H}{|}}{\underset{\underset{\displaystyle OH}{|}}{C}}-H \rightarrow H-\overset{\overset{\displaystyle H}{|}}{C}=\overset{\overset{\displaystyle H}{|}}{C}-H + H_2O$$

Ethyl alcohol          Ethylene

Ethylene is a colorless, flammable gas and is used as an anesthetic.

The physical properties of the alkenes are similar to those of the alkanes. However, because of the unstable double bond, the alkenes are chemically more reactive. They form *addition* compounds with $H_2$, $Cl_2$, $Br_2$, HCl, HBr, $H_2SO_4$, etc.

$$H-\underset{\underset{\displaystyle }{|}}{\overset{\overset{\displaystyle H}{|}}{C}}=\underset{\underset{\displaystyle }{|}}{\overset{\overset{\displaystyle H}{|}}{C}}-H \ + \ H_2 \ \xrightarrow{\text{Ni}} \ H-\underset{\underset{\displaystyle H}{|}}{\overset{\overset{\displaystyle H}{|}}{C}}-\underset{\underset{\displaystyle H}{|}}{\overset{\overset{\displaystyle H}{|}}{C}}-H$$

Ethane

$$H-\overset{\overset{\displaystyle H}{|}}{C}=\overset{\overset{\displaystyle H}{|}}{C}-H \ + \ Br_2 \ \longrightarrow \ H-\underset{\underset{\displaystyle Br}{|}}{\overset{\overset{\displaystyle H}{|}}{C}}-\underset{\underset{\displaystyle Br}{|}}{\overset{\overset{\displaystyle H}{|}}{C}}-H$$

Ethylene bromide

$$H-\overset{\overset{\displaystyle H}{|}}{C}=\overset{\overset{\displaystyle H}{|}}{C}-H \ + \ HBr \ \longrightarrow \ H-\underset{\underset{\displaystyle H}{|}}{\overset{\overset{\displaystyle H}{|}}{C}}-\underset{\underset{\displaystyle Br}{|}}{\overset{\overset{\displaystyle H}{|}}{C}}-H$$

Ethyl bromide

## ALKYNES

The alkynes or acetylene hydrocarbons contain two carbon atoms joined by a triple bond and are therefore even more unsaturated than the alkenes. They have the general formula $C_nH_{2n-2}$ and the ending *yne*. However, the simple alkynes are more often named as derivatives of acetylene.

$$C_2H_2 \qquad H-C\equiv C-H$$

Acetylene or
ethyne

$$C_3H_4 \qquad H-\underset{\underset{\displaystyle H}{|}}{\overset{\overset{\displaystyle H}{|}}{C}}-C\equiv C-H$$

Methylacetylene or
propyne

Chemically, the alkynes are even more reactive than the alkenes and will undergo addition reactions with $H_2$, $Br_2$, HBr, etc.

Acetylene, a colorless, flammable gas, is the most important compound of the series. It is prepared by the action of water upon calcium carbide.

$$CaC_2 + 2H_2O \rightarrow H—C\equiv C—H + Ca(OH)_2$$

It is used in the oxyacetylene torch for cutting and welding of steel, and for organic synthesis.

# 14. HALOGEN DERIVATIVES, ALCOHOLS, ETHERS, ALDEHYDES, AND KETONES

Whenever one or more hydrogen atoms of a hydrocarbon are replaced by atoms or groups of atoms, a hydrocarbon derivative is formed. Alkyl halides, alcohols, aldehydes, ketones, organic acids, esters, ethers, and amines are all derivatives of the hydrocarbons.

## HALOGEN DERIVATIVES

Halogen derivatives are formed when one or more hydrogen atoms in a hydrocarbon are replaced by halogen atoms. *Alkyl halides* are monohalogen compounds having the general formula RX or $C_nH_{2n+1}X$, where R represents any alkyl radical and X any halogen.

| | |
|---|---|
| $CH_3Cl$ | Methyl chloride |
| $C_2H_5Br$ | Ethyl bromide |
| $C_3H_7I$ | Propyl iodide |

*Polyhalides* or polyhalogen derivatives contain two or more halogen atoms.

| | |
|---|---|
| $CH_2Cl_2$ | Methylene chloride |
| $CCl_4$ | Carbon tetrachloride |

**Important Halogen Derivatives.** Halogen derivatives of hydrocarbons do not occur in nature. However, they are very useful because a wide variety of organic compounds can be made from them. The following halogen derivatives are especially important in medical practice and industry.

ETHYL CHLORIDE, $C_2H_5Cl$. Ethyl chloride, a volatile liquid, is occasionally used as a local anesthetic in minor surgery, since its rapid evaporation on the skin numbs the nerve endings.

CHLOROFORM, $CHCl_3$. Chloroform is a volatile, sweet-smelling, nonflammable liquid. It was formerly used as a general anesthetic. It is also a solvent for fats and oils.

IODOFORM, $CHI_3$. Iodoform is a pale yellow solid with the

characteristic odor of iodine. It is an effective antiseptic, but it is not widely used because of its disagreeable and persistent odor.

CARBON TETRACHLORIDE, $CCl_4$. Carbon tetrachloride is a colorless, nonflammable, volatile liquid. It is used in dry cleaning and in the Pyrene type fire extinguisher.

DICHLORODIFLUOROMETHANE (FREON), $CCl_2F_2$. Dichlorodifluoromethane is the most widely used refrigerant in refrigerators and air conditioners. It is also used as a propellant in aerosol bombs.

ETHYLENE BROMIDE, $CH_2BrCH_2Br$. Ethylene bromide is used to counteract the undersirable effect of lead tetraethyl contained in most high-test gasolines.

TETRACHLOROETHYLENE, $CCl_2=CCl_2$. Tetrachloroethylene is used to treat hookworm.

Unsaturated hydrocarbons such as ethylene form halogen derivatives by *adding* halogen atoms directly. There is no replacement of hydrogen atoms and no by-products are formed.

$$CH_2=CH_2 + Br_2 \rightarrow \begin{matrix} & H & H \\ & | & | \\ H- & C- & C-H \\ & | & | \\ & Br & Br \end{matrix}$$

Ethylene                Ethylene bromide

## ALCOHOLS

Alcohols are hydrocarbon derivatives in which one or more hydrogen atoms have been replaced by the hydroxyl or OH group. Their general formula is R—OH. They are classified in two ways: according to the number of hydroxyl groups, and according to the position of the carbon atom to which the OH group is attached.

**Classification According to the Number of OH Groups.** Alcohols that contain one OH group are called *monohydric alcohols.* Those containing two OH groups are called *dihydric alcohols* and those containing three OH groups, *trihydric alcohols. Polyhydric alcohols* are alcohols having more than one OH group.

$$CH_3-OH \qquad \begin{matrix} CH_2-CH_2 \\ | \quad\quad | \\ OH \quad OH \end{matrix} \qquad \begin{matrix} CH_2-CH-CH_2 \\ | \quad\quad | \quad\quad | \\ OH \quad OH \quad OH \end{matrix}$$

Methyl alcohol        Ethylene glycol            Glycerol
(monohydric)          (dihydric)                (trihydric)

**Classification According to the Position of the Carbon Atom to Which the OH Group is Attached.**   *Primary alcohols* are alcohols in which the OH group is attached to a primary or end carbon. *Secondary alcohols* are alcohols in which the OH group is attached to a secondary carbon, a carbon which is attached to two other carbon atoms. In *tertiary alcohols* the OH group is attached to a tertiary carbon, which is a carbon attached to three other carbon atoms.

$$CH_3-CH_2-CH_2-OH \qquad CH_3-\underset{\underset{OH}{|}}{CH}-CH_3 \qquad CH_3-\underset{\underset{OH}{|}}{\overset{\overset{CH_3}{|}}{C}}-CH_3$$

n-Propyl alcohol      Isopropyl alcohol      Tertiary butyl alcohol

(primary)            (secondary)         (tertiary)

**Naming of Alcohols.**   Simple alcohols are commonly named as derivatives of the hydrocarbons from which they are formed; the word *alcohol* is added to the common name of the radical, as in methyl alcohol and ethyl alcohol.  However, the I.U.C. names, which are formed by adding the suffix *-ol* to the hydrocarbon root, are gaining in usage.  Thus, methyl alcohol becomes methanol and ethyl alcohol, ethanol.  Alcohols having two OH groups are commonly known as *glycols*, e.g., ethylene glycol.

**Physical Properties of Alcohols.**   The presence of the OH group in the alcohol makes possible hydrogen bonding or association, which is responsible for boiling points higher than those of the corresponding hydrocarbons. It also accounts for the water solubility of the lower members.

**Chemical Properties of Alcohols.**   Because alcohols are covalent compounds and do not ionize, their reactions are much slower than those of inorganic hydroxides.  Alcohols burn, forming carbon dioxide and water with the production of much heat.

$$CH_3-CH_2-OH + 3O_2 \rightarrow 2CO_2 + 3H_2O$$
Ethyl alcohol

Primary alcohols are partially oxidized to aldehydes.

$$H-\underset{\underset{H}{|}}{\overset{\overset{H}{|}}{C}}-\underset{\underset{H}{|}}{\overset{\overset{H}{|}}{C}}-OH + [O] \rightarrow H-\underset{\underset{H}{|}}{\overset{\overset{H}{|}}{C}}-\overset{\overset{H}{|}}{C}=O + H_2O$$

Ethyl alcohol            Acetaldehyde

Secondary alcohols are partially oxidized to ketones.

$$H-\underset{\underset{H}{|}}{\overset{\overset{H}{|}}{C}}-\underset{\underset{H}{|}}{\overset{\overset{OH}{|}}{C}}-\underset{\underset{H}{|}}{\overset{\overset{H}{|}}{C}}-H + [O] \rightarrow H-\underset{\underset{H}{|}}{\overset{\overset{H}{|}}{C}}-\overset{\overset{O}{\|}}{C}-\underset{\underset{H}{|}}{\overset{\overset{H}{|}}{C}}-H + H_2O$$

Isopropyl alcohol                    Acetone

Alcohols react with organic acids to form esters.

$$CH_3COOH + C_2H_5OH \rightarrow CH_3COOC_2H_5 + H_2O$$

Acetic acid          Ethyl          Ethyl acetate
                    alcohol

Alcohols react slowly with sodium, which replaces the hydrogen in the hydroxyl group.

$$2CH_3OH + 2Na \rightarrow H_2 + 2CH_3ONa$$

Methyl                          Sodium
alcohol                        methoxide

**Important Alcohols.**   The most important alcohols are methanol, ethanol, ethylene glycol, and glycerol.

METHANOL (METHYL ALCOHOL), $CH_3OH$.   Methanol is also known as wood alcohol because it was formerly made exclusively by the destructive distillation of wood.   Most of the methanol produced today is made synthetically from hydrogen and carbon monoxide in the presence of a suitable catalyst and at high temperature and pressure.

$$2H_2 + CO \rightarrow CH_3OH$$

Methanol is a colorless liquid with a characteristic odor, soluble in water in all proportions.   It is poisonous, small doses producing blindness, large doses causing death.   It is used as a solvent for shellacs and varnishes, as an antifreeze for automobile radiators, as a denaturant for ethyl alcohol, and as raw material for conversion to formaldehyde.

ETHANOL (ETHYL ALCOHOL), $C_2H_5OH$.   Ethanol is produced by the fermentation of the sugar in molasses or of the starches in various grains (hence its name, "grain alcohol").   This process is made possible by means of enzymes (organic catalysts) present in sprouted barley and yeast.

$$C_{12}H_{22}O_{11} + H_2O \xrightarrow{\text{enzyme}} 2C_6H_{12}O_6$$

(complex      Glucose
sugar)       (simple
         sugar)

$$C_6H_{12}O_6 \xrightarrow{\text{enzyme}} 2C_2H_5OH + 2CO_2$$

Glucose    Ethyl  Carbon
      alcohol dioxide

Ethanol is a colorless liquid with a characteristic pleasant odor and is soluble in water in all proportions. It is used in intoxicating beverages and as a solvent, fuel, antiseptic, an antifreeze in automobile radiators, and as a raw material for making ether, chloroform, etc.

*Denatured alcohol* is ethyl alcohol to which such poisonous or ill-smelling substances as methyl alcohol, pyridine, or benzene have been added to render it unfit for drinking purposes.

ETHYLENE GLYCOL, $CH_2OH—CH_2OH$. Ethylene glycol, a dihydric alcohol, is prepared from ethylene through the formation of ethylene oxide or ethylene chlorohydrin. It is a colorless, water-soluble, poisonous liquid with a sweet taste. It is used as a solvent and as a permanent antifreeze in automobile radiators, sold under such trade names as Prestone and Zerex.

GLYCEROL (GLYCERINE), $CH_2OH—CHOH—CH_2OH$. Glycerol, a trihydric alcohol, is obtained as a by-product of the manufacture of soaps from animal fats and vegetable oils. It is a colorless, viscous, hygroscopic, sweet-tasting liquid, soluble in water in all proportions. Glycerol is used in making cosmetics, lotions, and explosives (nitroglycerine and dynamite).

## ETHERS

Ethers may be considered as organic oxides. They have two alkyl radicals joined to an oxygen atom and their general formula is R—O—R'. If the two alkyl radicals are the same, the ether is a *simple ether;* if they are different, the ether is a *mixed ether.*

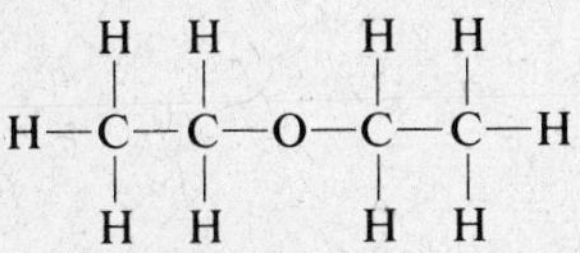

Diethyl ether      Methyl ethyl ether
(a simple ether)     (a mixed ether)

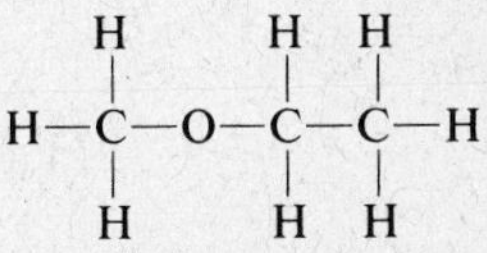

Ethers are commonly named by giving the names of the alkyl groups attached to the oxygen, followed by the word *ether*. In the I.U.C. system, the *-yl* of the alkyl radical is replaced by *-oxy*. The position of the oxygen atom is indicated by a number prefix.

$$CH_3-O-CH_2-CH_2-CH_3$$
*n*-Propyl methyl ether

$$CH_3-CH_2-CH_2-\underset{\underset{\displaystyle OCH_3}{|}}{CH}-CH_2-CH_3$$

3-Methoxyhexane

DIETHYL ETHER (ETHYL ETHER OR ETHER), $(C_2H_5)_2O$. Ether is prepared by the dehydrating action of concentrated sulfuric acid on ethyl alcohol.

$$C_2H_5-OH + HO-C_2H_5 \xrightarrow{H_2SO_4 \text{ (conc.)}} C_2H_5-O-C_2H_5 + H_2O$$
Ethyl alcohol                                      Diethyl ether

It is a light, volatile, flammable liquid, insoluble in water and relatively unreactive. It is used as a solvent for fats and as a general anesthetic.

DIVINYL ETHER (VINETHENE), $CH_2=CH-O-CH=CH_2$. This unsaturated ether is also used as a general anesthetic. It readily forms an explosive mixture with air.

## ALDEHYDES AND KETONES

Aldehydes and ketones may be considered as derivatives of hydrocarbons in which two hydrogen atoms on the same carbon atom have been replaced by an oxygen atom to form the carbonyl group (C=O). In aldehydes the carbonyl group is attached to one carbon atom and one hydrogen atom ($R-\overset{\overset{\displaystyle O}{||}}{C}-H$); in ketones it is attached to two carbon atoms ($R-\overset{\overset{\displaystyle O}{||}}{C}-R'$). If the two R's are the same, it is a simple ketone; if they are different, it is a mixed ketone.

$$CH_3-CH_2-\underset{}{\overset{\overset{\displaystyle H}{|}}{C}}=O \qquad CH_3-\overset{\overset{\displaystyle O}{||}}{C}-CH_3 \qquad CH_3-\overset{\overset{\displaystyle O}{||}}{C}-C_2H_5$$

Propionaldehyde          Dimethyl ketone          Methyl ethyl ketone
                         (a simple ketone)        (a mixed ketone)

**Naming of Aldehydes.**  Aldehydes are commonly named from the carboxylic acids they form when oxidized.  Thus the aldehyde that forms acetic acid is called acetaldehyde.  In the I.U.C. system, the -*e* ending of the longest chain hydrocarbon that includes the -CHO group is replaced by -*al*.  Thus, acetaldehyde ($CH_3$—CHO) is called ethanal.

**Naming of Ketones.**  Simple ketones are named by giving the names of the radicals attached to the carbonyl group, followed by the word *ketone*.  The simplest member of the ketone series is acetone ($CH_3$—CO—$CH_3$).  In the I.U.C. system, the -*e* ending of the longest chain hydrocarbon that includes the carbonyl group is replaced by -*one*.  The position of the carbonyl group is indicated by a number prefix.

$$CH_3—CO—CH_2CH_3 \qquad CH_3—CO—CH_2—CH_2—CH_3$$

Methyl ethyl ketone  ·  2-Pentanone

**Preparation of Aldehydes and Ketones.**  Aldehydes are prepared by the oxidation of primary alcohols, and ketones by the oxidation of secondary alcohols.

$$CH_3—CH_2—OH + [O] \rightarrow CH_3—CHO + H_2O$$

Ethyl alcohol          Acetaldehyde
(ethanol)              (ethanal)

$$CH_3—CHOH—CH_3 + [O] \rightarrow CH_3—CO—CH_3 + H_2O$$

Isopropyl alcohol                Acetone

**Chemical Properties of Aldehydes and Ketones.**  Although aldehydes and ketones have many properties in common because of the carbonyl group, aldehydes are more easily oxidized to acids and are good reducing agents.

$$\begin{array}{ccc} H & H & \\ | & | & \\ H—C—C{=}O & + [O] \rightarrow & H—C—C—OH \\ | & & | \\ H & & H \end{array}$$

Acetaldehyde                Acetic acid

Aldehydes reduce the solubilized cupric hydroxide in Fehling's and Benedict's solutions to an insoluble red precipitate of cuprous oxide.  Because simple sugars contain an aldehyde group, sugar in the urine can be detected by these tests.

$$R—C{=}O + 2Cu(OH)_2 \rightarrow R—C—OH + Cu_2O\downarrow + 2H_2O$$

Aldehyde     Reagent              Acid     Cuprous
                                           oxide

Aldehydes also reduce the silver hydroxide solubilized by ammonia in Tollens' reagent to a silver mirror deposited on the sides of the container.

$$\underset{\text{Aldehyde}}{R-\overset{\overset{\displaystyle H}{|}}{C}=O} + \underset{\text{Reagent}}{2AgOH} \rightarrow \underset{\text{Acid}}{R-\overset{\overset{\displaystyle O}{\|}}{C}-OH} + \underset{\substack{\text{Silver} \\ \text{mirror}}}{2Ag\downarrow} + H_2O$$

Aldehydes also polymerize. In *polymerization*, molecules unite with each other to form large molecules, or polymers.

**Important Aldehydes and Ketones.** The most important aldehydes and ketones are formaldehyde, acetaldehyde, and acetone.

FORMALDEHYDE, $H-\overset{\overset{\displaystyle H}{|}}{C}=O$. Formaldehyde is prepared by air oxidation of methyl alcohol over heated copper gauze.

$$2CH_3OH + O_2 \rightarrow 2HCHO + 2H_2O$$

It is a colorless gas with a penetrating odor, soluble in water. A 40 percent aqueous solution is known as formalin. Formaldehyde is used as a germicide, disinfectant, preservative for biological specimens, and embalming agent; also for leather tanning and plastic manufacture (Bakelite, Melmac). Formaldehyde reacts with ammonia to form hexamethylenetetramine (Urotropin), $(CH_2)_6N_4$, which is used as a urinary antiseptic.

ACETALDEHYDE, $CH_3-\overset{\overset{\displaystyle H}{|}}{C}=O$. Acetaldehyde is prepared by air oxidation of ethyl alcohol over heated copper guaze, or by the hydration of acetylene in the presence of mercuric sulfate. In the presence of a mineral acid, acetaldehyde polymerizes to *paraldehyde*, $(CH_3CHO)_3$, which is used as a sedative and hypnotic (sleep-producer). *Chloral*, $CCl_3CHO$, is another hypnotic prepared from acetaldehyde. It combines with water to form a colorless crystalline solid known as chloral hydrate.

ACETONE (DIMETHYL KETONE), $CH_3-\overset{\overset{\displaystyle O}{\|}}{C}-CH_3$. Acetone is prepared along with butyl alcohol by the Weizmann bacterial fermentation of cornstarch, by the destructive distillation of wood, by heating calcium acetate, or from the oxidation of isopropyl alcohol. Acetone is a colorless, combustible, low-boiling liquid, soluble in water in all proportions, and has a sweetish odor. It is used as a solvent for fats, gums, lacquers, resins, and plastics, and in the manufacture of chloroform, iodoform, dyes, acetate rayon, and explosives. It is found in excessive amounts in the urine and blood of untreated diabetics.

# 15. ORGANIC ACIDS, ORGANIC SALTS, ESTERS, AMIDES, AND AMINES

This chapter includes organic acids, which are another large group of hydrocarbon derivatives, and organic salts, esters, and amides which in turn may be derived from organic acids. Amines, which may be considered as organic derivatives of ammonia, are also discussed.

## ORGANIC ACIDS

Organic acids may be considered as derivatives of the hydrocarbons in which one or more of the hydrogen atoms have been replaced by a carboxyl group, $\overset{\overset{O}{\|}}{C}$—OH. Their general formula is R—COOH. This applies to the *monocarboxylic*, or monobasic, acids which are those containing only one carboxyl group. Organic acids containing two carboxyl groups are called *dicarboxylic acids* and those containing three carboxyl groups, *tricarboxylic acids*.

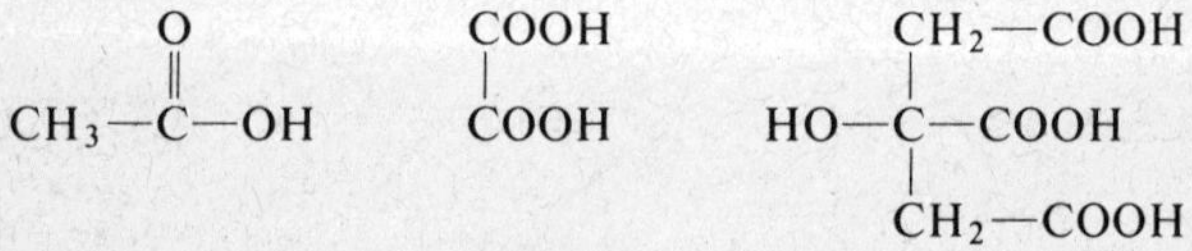

$$CH_3-\overset{\overset{O}{\|}}{C}-OH \qquad \begin{matrix} COOH \\ | \\ COOH \end{matrix} \qquad \begin{matrix} CH_2-COOH \\ | \\ HO-C-COOH \\ | \\ CH_2-COOH \end{matrix}$$

| Acetic acid | Oxalic acid | Citric acid |
|---|---|---|
| (monocarboxylic) | (dicarboxylic) | (tricarboxylic) |

**Naming of Organic Acids.** Most of the simpler carboxylic acids are known by their common names. The I.U.C. system is used to name the more complex acids. The *-e* ending of the longest chain hydrocarbon containing the carboxyl group is replaced by *-oic* followed by the word *acid*.

| | Common Name | I.U.C. Name |
|---|---|---|
| HCOOH | Formic acid | Methanoic acid |
| $CH_3COOH$ | Acetic acid | Ethanoic acid |
| $CH_3CH_2COOH$ | Propionic acid | Propanoic acid |
| $CH_3CH_2CH_2COOH$ | Butyric acid | Butanoic acid |
| $CH_3CH_2CH_2CH_2COOH$ | Valeric acid | Pentanoic acid |

**Compounds Formed from Acids.** Compounds derived from acids by changes in the carboxyl group are called *acid derivatives*. Examples are given below.

Salts: 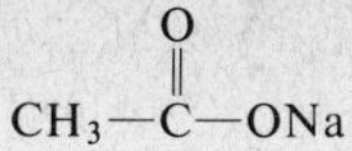
$$CH_3-\overset{\overset{\displaystyle O}{\|}}{C}-ONa$$
Sodium acetate

Esters: 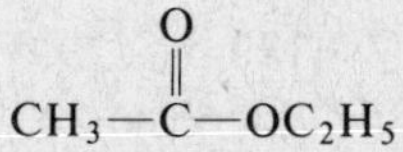
$$CH_3-\overset{\overset{\displaystyle O}{\|}}{C}-OC_2H_5$$
Ethyl acetate

Acid anhydrides: 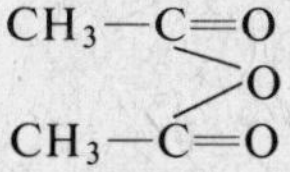
$$CH_3-C=O$$
$$\qquad\qquad >O$$
$$CH_3-C=O$$
Acetic anhydride

Acid halides: 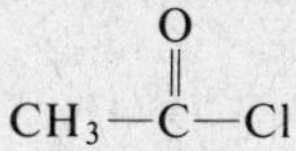
$$CH_3-\overset{\overset{\displaystyle O}{\|}}{C}-Cl$$
Acetyl chloride

Acid amides: 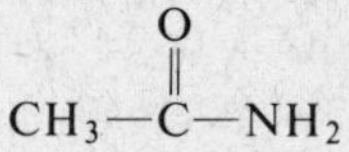
$$CH_3-\overset{\overset{\displaystyle O}{\|}}{C}-NH_2$$
Acetamide

Compounds formed from acids by changes in the alkyl group are called *substituted acids*.

Halogen-substituted acids: 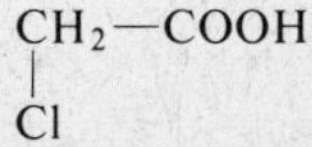
$$CH_2-COOH$$
$$|$$
$$Cl$$
Chloroacetic acid

Hydroxy acids: 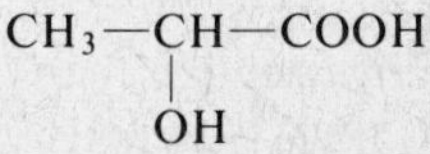
$$CH_3-CH-COOH$$
$$\qquad\quad |$$
$$\qquad\quad OH$$
Lactic acid
($\alpha$-hydroxypropionic acid)

Amino acids: 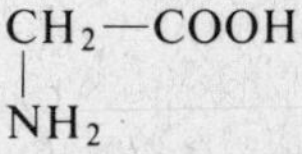
$$CH_2-COOH$$
$$|$$
$$NH_2$$
Glycine
(amino-acetic acid)

**Preparation and Properties of Organic Acids.** Organic acids may be prepared by the oxidation of primary alcohols or aldehydes, and from acid salts to which sulfuric acid has been added. Because they ionize only slightly, organic acids are weakly acidic. Acids in which the carboxyl group is attached to an alkyl radical are called *fatty acids* because some of them are obtained from animal and vegetable fats. They form a series whose lower members are liquids which are soluble in water. As the chain becomes longer the acids become less soluble. Those above $C_9$ are solid at room temperature and have no odor.

**Important Organic Acids.**    Some of the most important organic acids are formic, acetic, oxalic, lactic, tartaric, and citric.

FORMIC ACID, HCOOH.    Formic acid is a colorless liquid with a sharp, irritating odor, soluble in water in all proportions. It is present in the bite and sting of ants, bees, and some other insects; also in certain nettles. Formic acid is prepared by heating sodium hydroxide with carbon monoxide under pressure and treating the sodium salt formed with sulfuric acid. It is used as a reagent and in synthesizing organic compounds.

ACETIC ACID, $CH_3COOH$.    Acetic acid is a colorless liquid with a sharp, penetrating odor, soluble in water in all proportions. It is prepared by the destructive distillation of wood, the oxidation of acetaldehyde from acetylene, and the fermentation of fruit juices. Vinegar contains 4 percent acetic acid. Pure (100 percent) acetic acid is called *glacial acetic acid* because it freezes to an ice-like solid at 16.6°C. Acetic acid is used for making white vinegar, white lead, esters, and cellulose acetate; also in dyeing textiles and as a valuable laboratory reagent.

OXALIC ACID, HOOC—COOH.    Oxalic acid, a white crystalline solid, is present in rhubarb and several other plants. It is a poison since it removes calcium from the body. Oxalic acid is used to bleach straw and to remove iron rust and ink stains; also in calico printing and in the textile dyeing and tanning industries.

LACTIC ACID, $CH_3$—CHOH—COOH.    Lactic acid is present in sour milk and sauerkraut. It forms in muscle tissue during activity. It is used in the manufacture of yeast and cheese and in the dyeing and leather industries.

TARTARIC ACID, HOOC—$(CHOH)_2$—COOH.    Tartaric acid and its potassium salt (cream of tartar) are present in grapes. It is used in soft drinks, in the dyeing industry, and as a mild cathartic.

CITRIC ACID,    $HOOCCH_2$—C(OH)(COOH)—$CH_2COOH$. Citric acid is present in all citrus fruits. It is prepared by the fermentation of starch and molasses. It is used in soft drinks and in the dyeing industry.

## ORGANIC SALTS

Organic salts are compounds formed from acids in which the hydrogen atom of the carboxyl group has been replaced by a metal or by the ammonium radical.

$$CH_3COOH + NaOH \rightarrow CH_3COONa + H_2O$$

Acetic acid        Sodium acetate

Organic salts are solids. Organic salts of alkali metals and ammonium salts are soluble in water and insoluble in ether. In aqueous solutions they ionize and undergo hydrolysis. Organic salts release their corresponding acids when they are treated with a mineral acid.

Some important organic salts, their formulas and uses, are listed in Table 15.1.

### TABLE 15.1    SOME IMPORTANT ORGANIC SALTS

| Salt | Formula | Uses |
|------|---------|------|
| Aluminum acetate | $(CH_3COO)_3Al$ | Astringent, antiseptic, internal and external disinfectant; in textile dyeing |
| Calcium lactate | $(C_3H_5O_3)_2Ca$ | Supplements calcium in diet |
| Calcium propionate | $(C_2H_5COO)_2Ca$ | Prevents molding in bread |
| Lead acetate | $(CH_3COO)_2Pb$ | Astringent; in dyeing and printing cottons |
| Magnesium citrate | $[C_3H_4OH(COO)_3]_2Mg_3$ | Saline cathartic |
| Potassium bitartrate (cream of tartar) | $KHC_4H_4O_6$ | Laxative, diuretic, baking powder, source of tartaric acid |
| Potassium oxalate | $(COOK)_2$ | Bleaches straw, prevents clotting of blood for analysis |
| Potassium sodium tartrate (Rochelle salt) | $KNaC_4H_4O_6$ | Saline cathartic, diuretic, Seidlitz powder, Fehling's solution |
| Sodium acetate | $CH_3COONa$ | Buffer, diuretic, mordant |
| Sodium citrate | $C_3H_4OH(COO)_3Na_3$ | Prevents clotting of blood in transfusions |
| Zinc stearate | $(C_{17}H_{35}COO)_2Zn$ | Astringent, dusting powder for infants |

## ESTERS

Esters may be considered as derivatives of acids in which the acid hydrogen has been replaced by an alkyl radical. Generally they are prepared by *esterification* of an acid with an alcohol, a reaction somewhat similar to neutralization. If the acid used is an organic acid, the ester is called an organic ester, which has the general formula R—CO—O—R'. If the acid used is an inorganic acid, the ester formed is called an inorganic ester. When the word ester is used without modification, it refers to an organic ester.

$$CH_3-\overset{O}{\overset{\|}{C}}-OH + C_2H_5OH \rightarrow CH_3-\overset{O}{\overset{\|}{C}}\ OC_2H_5 + H_2O$$

Acetic acid          Ethyl          Ethyl acetate
(an organic acid)    alcohol        (an organic ester)

$$HONO + C_2H_5OH \rightarrow C_2H_5ONO + H_2O$$

Nitrous acid    Ethyl      Ethyl nitrite
(an inorganic   alcohol    (an inorganic
acid)                      ester)

Esters are named from the carboxylic acid from which they were derived by changing the ending *-ic* to *-ate* and dropping the word *acid*. This is preceded by the name of the R' group. Thus, ethyl acetate, $CH_3COOCH_2CH_3$, is derived from acetic acid, $CH_3COOH$.

Because of the lack of association, esters have lower boiling points than both the alcohol and the acid of equal molecular weights. Except for a few lower members, they are insoluble in water but soluble in organic solvents.

The esters have a very pleasant fruity odor and are responsible for the flavor and fragrance of many fruits and flowers. Some esters and their odors are: amyl acetate—pear; isoamyl acetate—banana; amyl butyrate—apricot; ethyl butyrate—pineapple; isobutyl acetate—raspberry.

Esters undergo *hydrolysis* to form acids and alcohols. The hydrolysis may be catalyzed by either acids or alkalis. Alkaline hydrolysis is usually called *saponification*, a reaction also used in the preparation of soaps.

Hydrolysis:

$$CH_3-\overset{O}{\overset{\|}{C}}-OC_2H_5 + HOH \rightarrow CH_3-\overset{O}{\overset{\|}{C}}-OH + C_2H_5OH$$

Ethyl acetate                    Acetic acid      Ethyl
                                                  alcohol

Saponification:

$$CH_3-\overset{\overset{\displaystyle O}{\|}}{C}-OC_2H_5 + NaOH \rightarrow CH_3-\overset{\overset{\displaystyle O}{\|}}{C}-ONa + C_2H_5OH$$

Ethyl acetate            Sodium acetate     Ethyl alcohol

TABLE 15.2. SOME IMPORTANT ESTERS

| Ester | Formula | Uses |
|---|---|---|
| Butyl acetate | $CH_3COOC_4H_9$ | Solvent for proxylin lacquers |
| Isoamyl acetate (banana oil) | $CH_3COOCH_2CH_2CH(CH_3)_2$ | Paint drier, protective coating for tooth cavities. |
| Ethyl nitrite | $C_2H_5ONO$ | Its alcohol solution, "sweet spirit of niter," is a diuretic and vasodilator |
| Amyl nitrite | $C_5H_{11}ONO$ | Treatment of asthma and angina pectoris, vasodilator, antispasmodic |
| Glyceryl trinitrate (nitroglycerine) | $C_3H_5(NO_3)_3$ | Same as amyl nitrite; explosive |

# AMIDES

Amides are derivatives of organic acids (R—COOH) in which the OH group has been replaced by the amino ($NH_2$) group. Thus primary amides have the general formula R—CO—$NH_2$. They are named according to the acids from which they are derived by changing the -ic ending to -amide and dropping the word acid. Thus, acetamide, $CH_3CONH_2$, is derived from acetic acid, $CH_3COOH$.

Amides are usually crystalline solids with high melting points. The lower members of the series are hygroscopic, water-soluble compounds. Higher members are soluble in alcohol and ether.

**Important Amides.** Acetamide and urea are familiar amides.

ACETAMIDE, $CH_3CONH_2$. Acetamide is a typical amide. It may be prepared from ammonium acetate according to the following reaction:

$$CH_3-COONH_4 \overset{\Delta}{\rightarrow} CH_3-\overset{\overset{\displaystyle O}{\|}}{C}-OH + H-NH_2 \rightarrow$$

Ammonium acetate       Acetic acid      Ammonia

$$CH_3-\overset{\overset{\displaystyle O}{\|}}{C}-NH_2 + H_2O$$

Acetamide

Acetamide and other lower members of the series are excellent solvents for both organic and inorganic substances.

UREA, $H_2N-\overset{\overset{\displaystyle O}{\|}}{C}-NH_2$. Urea is the diamide of carbonic acid. It is the chief end-product of protein metabolism and is excreted in the urine. It was the first organic compound to be synthesized. Commercially it is prepared by heating carbon dioxide and ammonia under high pressure.

$$CO_2 + 2NH_3 \rightarrow C \overset{\displaystyle \nearrow NH_2}{\underset{\displaystyle \searrow NH_2}{=}O} + H_2O$$

Urea is used in the manufacture of fertilizers and plastics.

## AMINES

Amines are organic derivatives of ammonia in which one or more of the hydrogen atoms have been replaced by an alkyl radical. If one hydrogen is replaced, the amine formed is a *primary amine*; if two hydrogens are replaced, it is a *secondary amine*; and if all three hydrogens are replaced, it is a *tertiary amine*. Primary amines have the general formula $R-NH_2$.

$$\underset{\underset{\text{H}}{|}}{\text{H}-\text{N}-\text{H}} \qquad \underset{\underset{\text{H}}{|}}{\text{CH}_3-\text{N}-\text{H}} \qquad \underset{\underset{\text{CH}_3}{|}}{\text{CH}_3-\text{N}-\text{H}} \qquad \underset{\underset{\text{CH}_3}{|}}{\text{CH}_3-\text{N}-\text{CH}_3}$$

Ammonia     Methylamine     Dimethylamine     Trimethylamine
(primary)     (secondary)     (tertiary)

Amines are named by giving the alkyl groups on the nitrogen, followed by the ending *amine*.

Like ammonia, amines are basic and react with inorganic acids to form salts.

$$\underset{\underset{\text{H}}{|}}{\overset{\overset{\text{H}}{|}}{\text{CH}_3-\text{N}}} + \text{HCl} \rightarrow \left[\underset{\underset{\text{H}}{|}}{\overset{\overset{\text{H}}{|}}{\text{CH}_3-\text{N}-\text{H}}}\right]^{+} \text{Cl}^{-}$$

Methylamine        Methylamine
hydrochloride

Methylamine hydrochloride may be written as $CH_3NH_3{}^+Cl^-$ or $CH_3NH_2 \cdot HCl$.

*Quaternary ammonium salts* form when a tertiary amine reacts with an excess of alkyl halide. These salts are widely used in medicine.

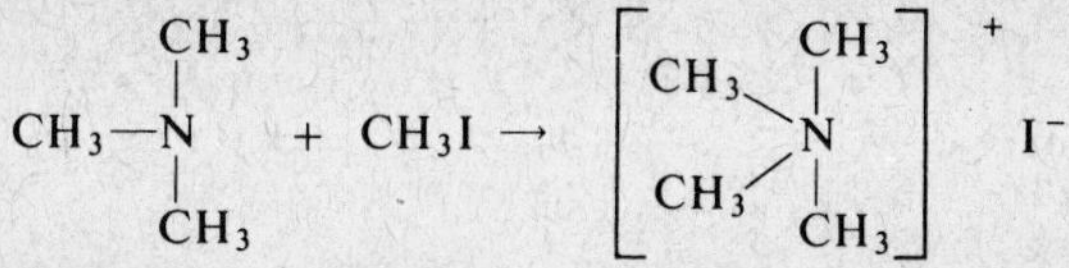

$$CH_3-\underset{\underset{CH_3}{|}}{\overset{\overset{CH_3}{|}}{N}} \;+\; CH_3I \;\longrightarrow\; \left[ \underset{\underset{CH_3}{|}}{\overset{\overset{CH_3}{|}}{\underset{CH_3}{\overset{CH_3}{N}}}} \right]^{+} \;\; I^{-}$$

| Trimethylamine | Methyl iodide | Tetramethylammonium iodide |

Low molecular weight amines are water soluble and have a characteristic fishy odor. Their vapors are flammable. Higher amines are less soluble and less odorous.

Some important amines, their formulas, and their occurrence and uses are listed in Table 15.3.

TABLE 15.3. SOME IMPORTANT AMINES

| Name | Formula | Occurrence and Uses |
|---|---|---|
| Methylamine | $CH_3NH_2$ | Used in dehairing animal hides |
| Dimethylamine | $(CH_3)_2NH$ | Used in dehairing animal hides |
| Trimethylamine | $(CH_3)_3N$ | Occurs in herring brines and beet sugar residue |
| Butylamine | $C_4H_9NH_2$ | Used as antioxidants and corrosion inhibitors |
| Amylamine | $C_5H_{11}NH_2$ | Corrosion inhibitor |
| Ethanolamine | $HO-CH_2-CH_2-NH_2$ | Occurs in cephalin; used as an emulsifying agent |
| Choline | $HO-CH_2-CH_2-N(CH_3)_3(OH)$ | Occurs in lecithin |
| Acetylcholine | $CH_3-COO-CH_2-CH_2-N(CH_3)_3(OH)$ | Released at nerve endings during muscular contractions |
| Putrescine (a ptomaine) | $H_2N-(CH_2)_4-NH_2$ | Occurs in putrefied proteins |
| Cadaverine (a ptomaine) | $H_2N-(CH_2)_5-NH_2$ | Occurs in putrefied proteins |

# 16. CYCLIC ORGANIC COMPOUNDS

Cyclic organic compounds are those in which the carbon atoms form a ring structure. This is in contrast to the aliphatic compounds in which the carbon atoms are arranged in an open-chain or branched-chain structure.

There are three classes of cyclic compounds: the alicyclic, the aromatic, and the heterocyclic compounds.

## ALICYCLIC COMPOUNDS

Alicyclic compounds are cyclic compounds corresponding to the straight-chain aliphatic compounds. Thus there are cycloalkanes, cycloalkenes, and cycloalkynes.

**Cycloalkanes.** The cycloalkanes, also called cycloparaffins, resemble the alkanes in that they are saturated, containing only single bonds. They differ from the alkanes in that their carbon atoms are arranged in a closed chain or ring structure. The most important simple cycloalkanes are the following:

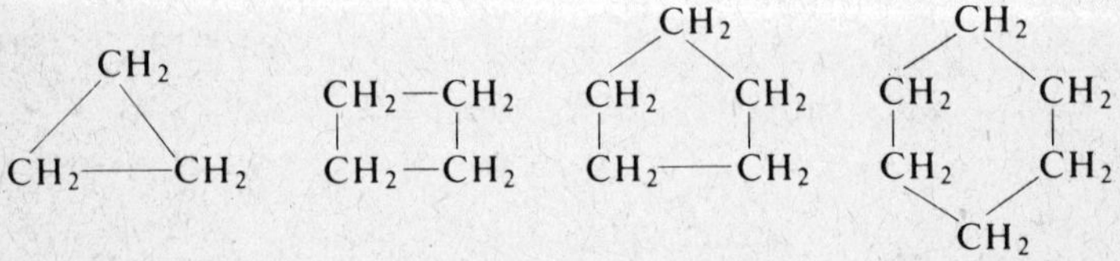

| Cyclopropane | Cyclobutane | Cyclopentane | Cyclohexane |

*Cyclopropane* is a sweet-smelling, colorless gas. It has been used as a general anesthetic since 1930.

## AROMATIC COMPOUNDS

Aromatic compounds are those organic compounds containing at least one closed ring of six carbon atoms joined by alternating single and double bonds. They are known as aromatic compounds because the first ones described had a pleasant aroma. However, not all compounds of this structure have odors or pleasant ones, nor is this an exclusive property of this group. Some aliphatic esters are also fragrant-smelling.

**Benzene.** Benzene, $C_6H_6$, is the parent aromatic compound, composed of six carbon and six hydrogen atoms in a closed ring with alternate double and single bonds. It may be represented by the following formula:

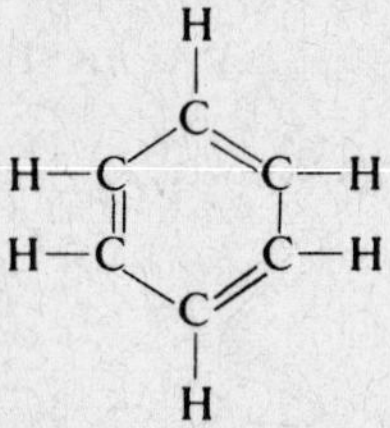

There is only one monosubstitution product of benzene because the six replaceable hydrogen atoms are equivalent. Disubstituted isomers are differentiated by using the prefixes *ortho* (*o*-), *meta* (*m*-), and *para* (*p*-).

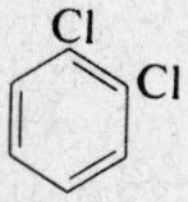

*o*-Dichlorobenzene 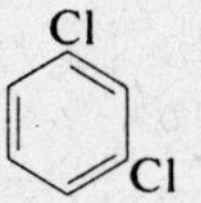      *m*-Dichlorobenzene 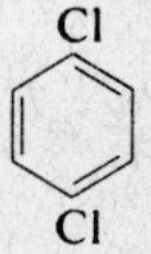      *p*-Dichlorobenzene

Tri or tetra substituted compounds are best named by using numerals.

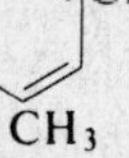

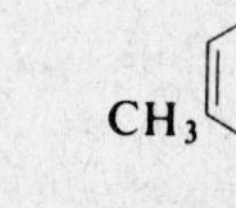

1,2,3-Trimethylbenzene

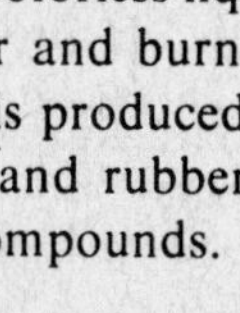

1,2,4-Trimethylbenzene 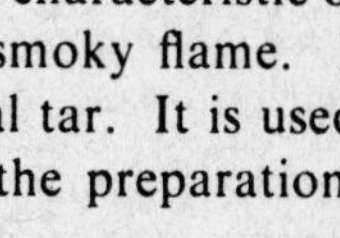      1,3,5-Trimethylbenzene

Benzene is a colorless liquid with a characteristic odor. It is insoluble in water and burns with a smoky flame. It is toxic to man. Benzene is produced from coal tar. It is used as a solvent for fats, resins, and rubber, and in the preparation of dyes and other organic compounds.

REACTIONS OF BENZENE.  In spite of the presence of double bonds in the molecule, benzene acts more like a saturated than like an unsaturated compound because most of its reactions are substitution rather than addition.

*Halogenation (chlorination, bromination)*

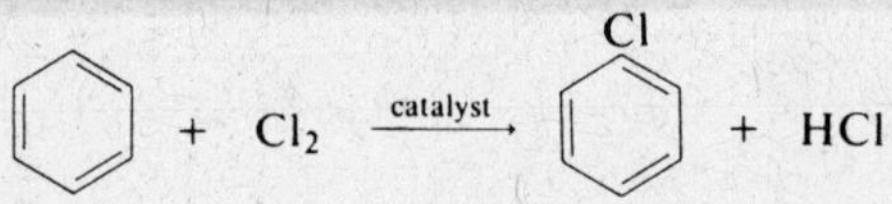

Benzene    $+ Cl_2 \xrightarrow{\text{catalyst}}$    Chlorobenzene *    $+ HCl$

*Nitration*

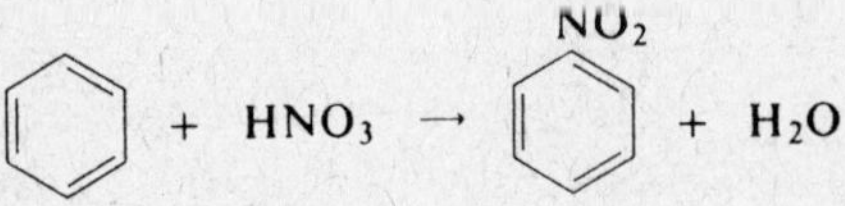

$+ HNO_3 \rightarrow$    Nitrobenzene    $+ H_2O$

*Sulfonation*

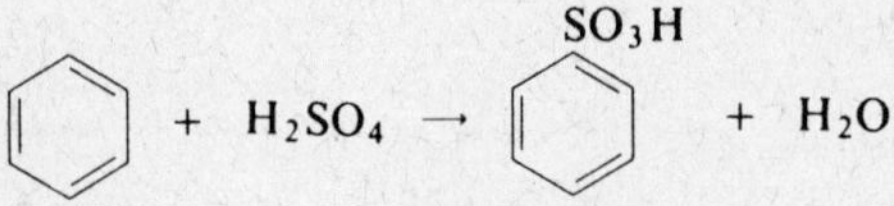

$+ H_2SO_4 \rightarrow$    Benzenesulfonic acid    $+ H_2O$

OTHER MEMBERS OF THE BENZENE SERIES.  Toluene, the xylenes, and ethylbenzene are also coal-tar products.

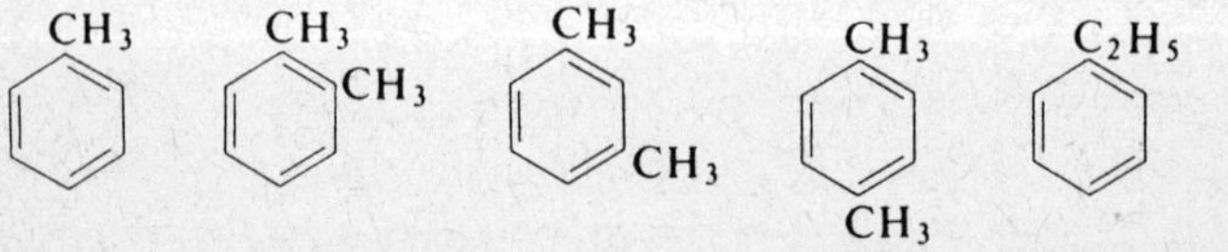

Toluene    *o*-Xylene    *m*-Xylene    *p*-Xylene    Ethylbenzene

*Toluene* is an important organic solvent and is used in the manufacture of benzoic acid, benzaldehyde, trinitrotoluene (TNT), dyes, and other organic compounds.  It is also used as a preservative for urine.  The *xylenes* are used for the preparation of perfumes, dyes, and explosives, and for cleaning the oil immersion lens of the microscope.  They are obtained from coal tar as a mixture called xylol.

**Polynuclear or Condensed Aromatic Hydrocarbons.**  Polynuclear hydrocarbons are those in which two or more benzene

---

*Chlorobenzene, $C_6H_5$—Cl, may be called phenyl chloride since $C_6H_5$—, which is benzene minus a hydrogen, is known as the *phenyl* radical.

rings are fused together. The most important hydrocarbons of this type are naphthalene, anthracene, and phenanthrene, all of which are obtained from coal tar.

NAPHTHALENE, $C_{10}H_8$. Naphthalene consists of two benzene rings fused together, with two carbon atoms common to each ring. It has the following structure:

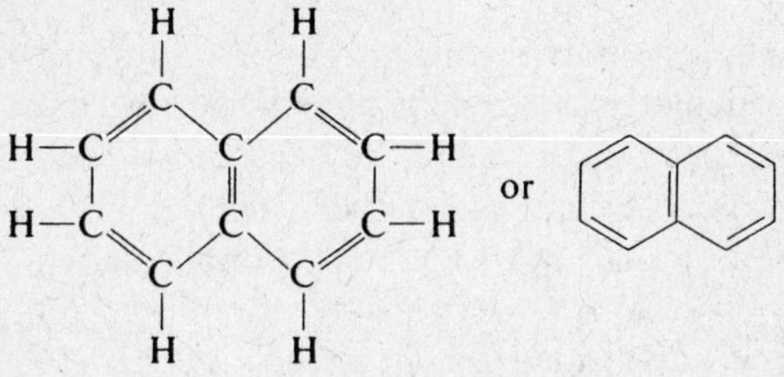

Naphthalene is used for making mothballs and in the manufacture of various dyes and dye intermediates. Most of the naphthalene produced is converted into phthalic anhydride for use in the preparation of polyester resins.

ANTHRACENE, $C_{14}H_{10}$. Anthracene consists of three fused benzene rings. It has the following structure:

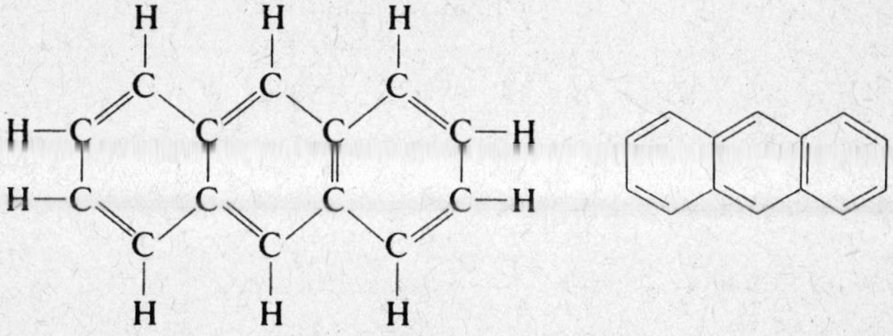

The most important derivative of anthracene is anthraquinone, which is used in the manufacture of anthraquinone dyes.

PHENANTHRENE, $C_{14}H_{10}$. Phenanthrene is an isomer of anthracene. Its structure is as follows:

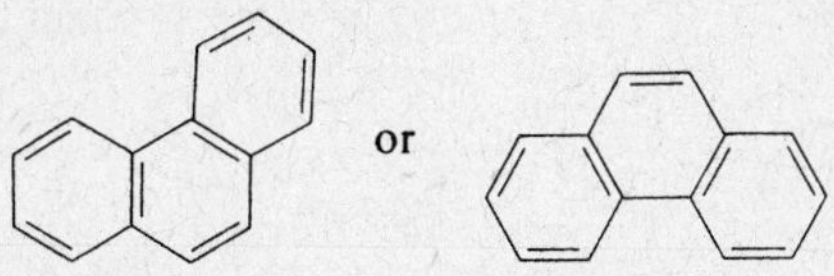

The phenanthrene ring is an integral part of the sterols, sex hormones, and vitamin D.

**Halogen Derivatives of Benzene.** The two most useful halogen derivatives of benzene are p-dichlorobenzene and DDT, whose structures are as follows:

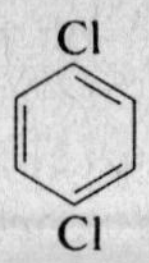

p-Dichlorobenzene 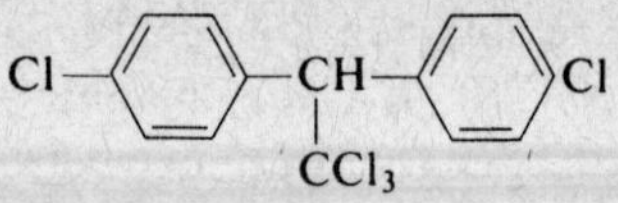                    DDT

p-Dichlorobenzene is replacing the once important naphthalene as a more efficient moth repellent. DDT, which is dichlorodiphenyltrichloroethane, is one of the best known insecticides.

Other halogen derivatives of benzene include Chloramine-T, halozone, and benzene hexachloride (Gammexane).

**Aromatic Amines and Derivatives.** Aromatic amines contain an amino or a substituted amino group attached to the benzene ring. Aromatic amines and their derivatives are among the most important organic compounds, for from them are synthesized organic compounds used as dyes and dye intermediates, indicators, rubber accelerators, photographic developers, antioxidants, and drugs. The names and structures of the more important aromatic amines and their derivatives are given below.

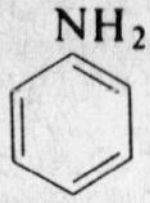

Aniline 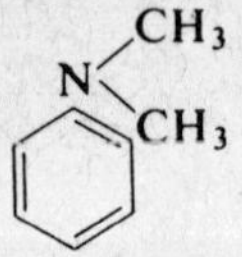      Dimethylaniline 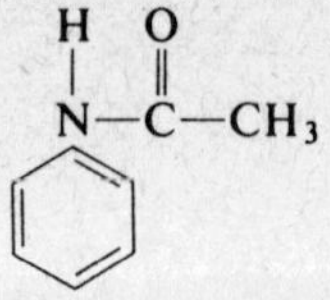      Acetanilide 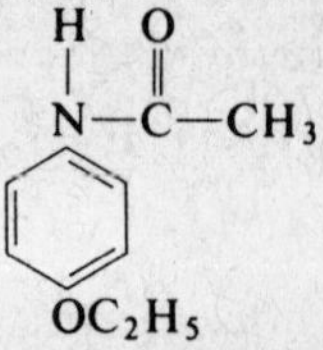      Phenacetin

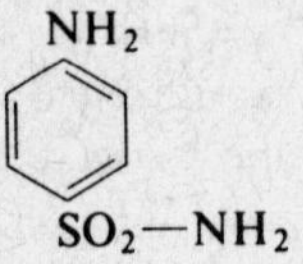

Sulfanilamide 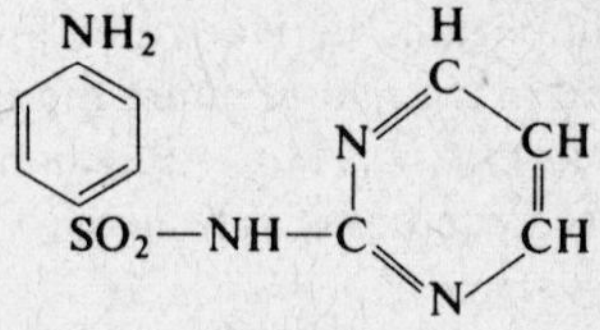            Sulfadiazine

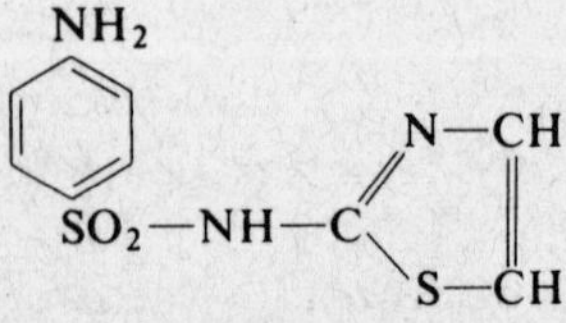

Sulfathiazole

ANILINE.   Aniline (aminobenzene), the parent aromatic amine, is a coal-tar product.  It is also prepared commercially by the reduction of nitrobenzene, using iron and water with a little ferrous sulfate.

$$NO_2 \xrightarrow{\text{reduction}} NH_2$$

Nitrobenzene          Aniline

Aniline is an oily liquid, slightly soluble in water and soluble in organic solvents.  It is colorless when freshly distilled, but darkens on exposure to air and light.  Like the aliphatic amines, it is basic and reacts with acids to form salts.  Because it is toxic, contact with the liquid or inhalation of the vapors should be avoided.

Aniline is used as a solvent and for the synthesis of other amines and other organic compounds.

OTHER AROMATIC AMINES AND DERIVATIVES.   *Dimethylaniline* is an important dye intermediate.  *Acetanilide*, prepared from aniline and acetyl chloride, is sometimes used as an antipyretic and analgesic, but it is somewhat toxic.  *Phenacetin* is used for the same purpose as acetanilide, but is less toxic.  *Sulfanilamide*, the forerunner of the sulfa drugs, is prepared from acetanilide and is an effective internal antiseptic against staphylococcus, streptococcus, pneumococcus, and gonococcus infections.  *Sulfadiazine*, *sulfathiazole*, and *sulfasoxazole* are less toxic but more effective than sulfanilamide in the treatment of certain types of infections.

**Aromatic Alcohols.**   Aromatic alcohols have an OH group attached to an aliphatic side chain.  Their chemical properties are similar to those of the aliphatic alcohols.  Most of the aromatic alcohols have pleasant odors and are therefore used in the manufacture of perfumes.  Two aromatic alcohols that have medicinal uses are benzyl alcohol and ephedrine.

Benzyl alcohol          Ephedrine

Benzyl alcohol is used as a local anesthetic and in lotions and ointments to relieve itching.  Ephedrine, obtained from the Chinese herb *Ma huang*, has the property of constricting the peripheral blood vessels and dilating the bronchioles; it is therefore used in the treatment of hay fever, asthma, sinusitis, and head colds.

**Phenols.** Phenols are aromatic compounds in which one or more of the hydrogen atoms on the benzene ring have been replaced by the hydroxyl group. Because the OH group is attached directly to the benzene ring, the properties of phenols are different from those of aromatic alcohols. Phenols are weakly acidic and react with bases to form salts called phenoxides. The structures of the most important phenols are as follows:

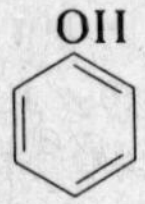

Phenol 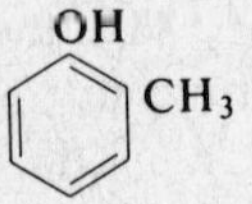     *o*-Cresol 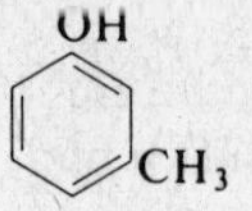     *m*-Cresol      *p*-Cresol

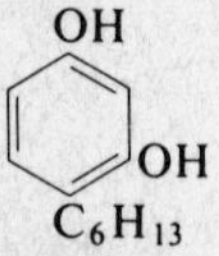

Hexylresorcinol      *p*-Aminophenol 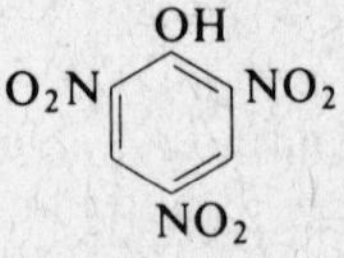     Picric acid

PHENOL. Phenol, or carbolic acid, is a coal-tar product but is prepared in large quantities by the alkaline hydrolysis of chlorobenzene and other methods.

Phenol is a white crystalline solid with a characteristic odor. On standing it turns pink to red as the result of oxidation. It is fairly soluble in water, yielding a weak acid solution with good antiseptic properties. It is very corrosive, causing blisters and deep burns when it comes in contact with the skin. Immediate washing with alcohol is recommended.

Phenol is used as a disinfectant, as a preservative for biological specimens, as a cautery, and in making the plastic Bakelite.

CRESOLS. Cresols, or methylphenols, are methyl derivatives of phenol. They are obtained from coal tar and are generally used as a mixture of *ortho-*, *meta-*, and *para*-cresols. Cresols are used for preserving wood and in the manufacture of synthetic resins. Lysol is made from cresols by emulsifying them with a soap solution. Cresols are more effective antiseptics than phenol and are not as toxic.

**OTHER PHENOLS.** *Hexylresorcinol* as an antiseptic is about 45 times as effective as phenol and is not injurious to body tissues. *p-Aminophenol* and its derivatives are mild reducing agents and are used as photographic developers. *Picric acid*, or 2,4,6-trinitrophenol, is a yellow crystalline solid prepared by heating phenol with a mixture of nitric and sulfuric acids. It is used as an explosive and as a dye for silk and wool.

**Aromatic Aldehydes.** In aromatic aldehydes the aldehyde group —CHO is attached either to one of the carbons in the benzene ring or to a carbon in a side chain. The structures of some important aromatic aldehydes are given below.

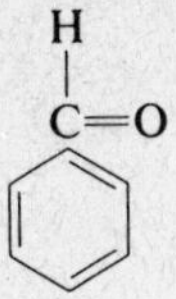

Benzaldehyde 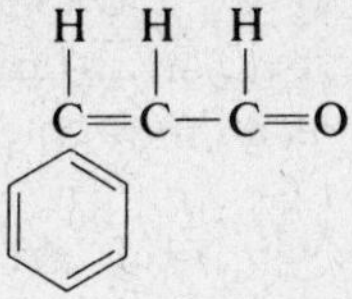          Cinnamaldehyde 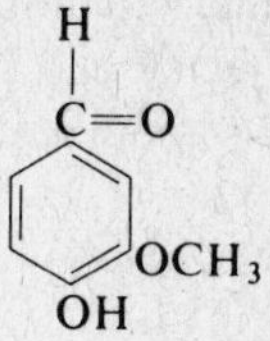          Vanillin

*Benzaldehyde*, a colorless oil with a pleasant almond-like odor, is a constituent of oil of bitter almonds. It is used in flavoring agents and perfumes, and in the preparation of drugs, dyes, and other organic compounds. *Cinnamaldehyde* has the odor and flavor of cinnamon and is used as a flavoring agent. *Vanillin*, the flavoring principle of the vanilla bean, is usually made by hydrolyzing lignin, a by-product of the paper industry.

**Aromatic Ketones.** In aromatic ketones the carbonyl group can be attached to two aromatic rings, as in benzophenone, or to an aromatic and an aliphatic group, as in acetophenone. Benzoin is a hydroxy ketone.

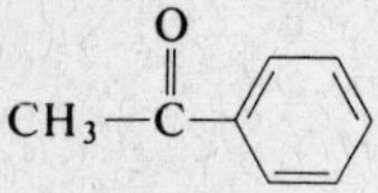

Acetophenone 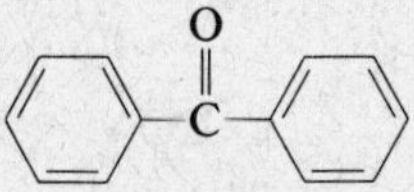          Benzophenone 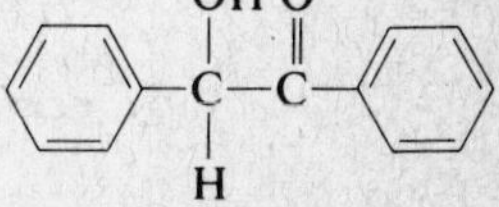          Benzoin

*Acetophenone*, a colorless liquid with a characteristic odor, is used in organic synthesis and as a hypnone. *Benzophenone* is a colorless solid with a pleasant odor. It is used in the preparation of perfumes, soaps, and other organic compounds. *Benzoin*, also a colorless solid, is used in the preparation of tincture of benzoin for treating ulcers and as a stimulating expectorant.

**Aromatic Acids.**    Aromatic acids are compounds in which one or more hydrogen atoms on the benzene ring have been replaced by carboxyl groups, or in which a carboxyl group is attached to a side chain.

BENZOIC ACID.    Benzoic acid and its salt, *sodium benzoate*, have the following structures:

COOH    COONa

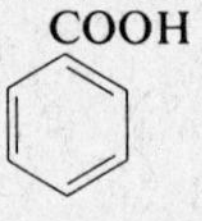

Benzoic acid 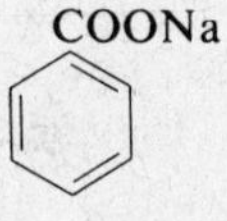        Sodium benzoate

Benzoic acid, a colorless crystalline solid, is found in cranberries and gum benzoin.  It is used as an expectorant for bronchitis. Sodium benzoate is used as an antiseptic and as a food preservative.

SALICYLIC ACID.    Salicylic acid is formed when a hydroxyl group is substituted on the *ortho* carbon of benzoic acid.  Salicylic acid and its important derivatives have the following structures:

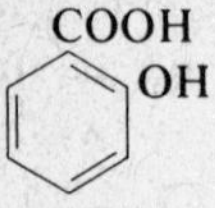

COOH    COONa    COOCH$_3$    COOH    O
   OH         OH            OH               O—C—CH$_3$

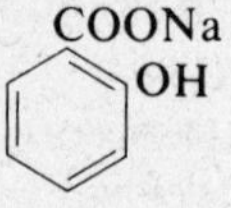

Salicylic      Sodium       Methyl          Acetylsalicylic
  acid        salicylate    salicylate           acid

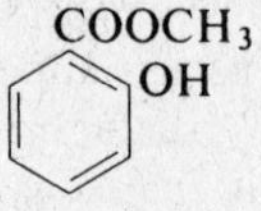

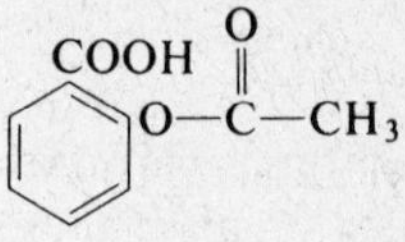

Salicylic acid is used as an antiseptic and in corn-removing agents. *Sodium salicylate* is used as an antipyretic (fever reducer) and in the treatment of rheumatism and arthritis.  *Methyl salicylate* (oil of wintergreen), an aromatic ester, is used as a flavoring agent and as a counterirritant in liniments and ointments.  *Acetylsalicylic acid* (aspirin) is a widely used antipyretic and analgesic agent.

## HETEROCYCLIC COMPOUNDS

The heterocyclic compounds are cyclic compounds that contain other elements besides carbon in the ring, in contrast to the homocyclic or carbocyclic compounds that contain only carbon atoms in the ring, such as benzene and its derivatives.  The majority of the heterocyclic compounds consist of five- or six-membered rings that contain either one or two elements other than carbon.  The three most common hetero elements found in these compounds are oxygen, nitrogen, and sulfur.  The most important parent ring systems and their derivatives are the following:

**Five-membered Rings with One Hetero Atom.** *Furan* contains an oxygen atom, *pyrrole* a nitrogen atom, and *thiophene* a sulfur atom in the ring. The pyrrole ring occurs in hemoglobin, chlorophyll, and bilirubin, a bile pigment.

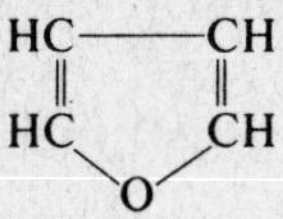

Furan 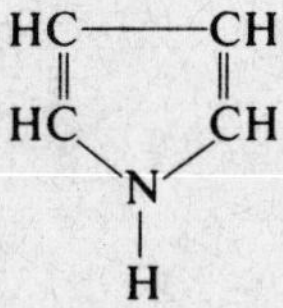　　　　Pyrrole 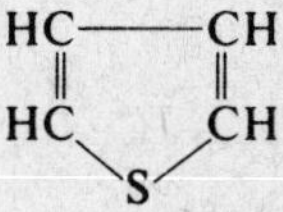　　　　Thiophene

**Five-membered Rings with Two Hetero Atoms.** *Thiazole* contains both nitrogen and sulfur in the ring. Penicillin is a substituted thiazole compound. *Imidazole* contains two nitrogen atoms in the ring. Histidine, an amino acid, contains the imidazole ring.

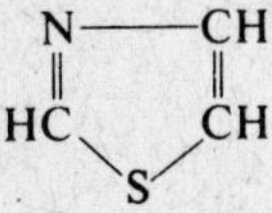

Thiazole 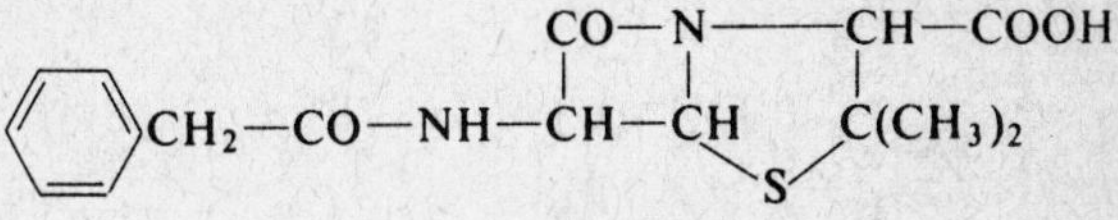　　　　　　　　Penicillin G

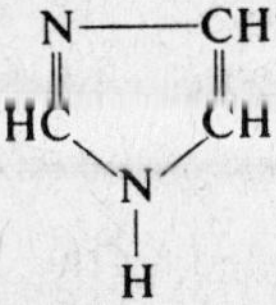

Imidazole 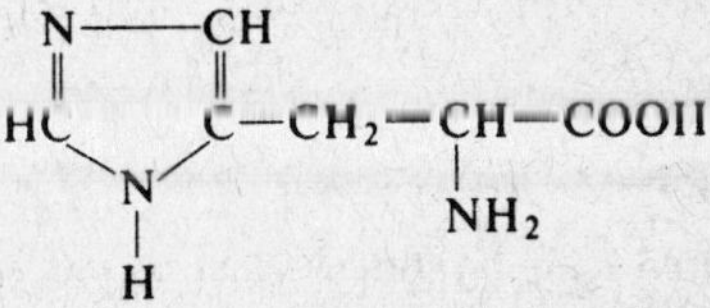　　　　Histidine
(an amino acid)

**Six-membered Rings with One Hetero Atom.** *Pyridine* has one nitrogen atom in the ring. Nicotinic acid and nicotinamide, which are members of the vitamin B complex, are derivatives of pyridine.

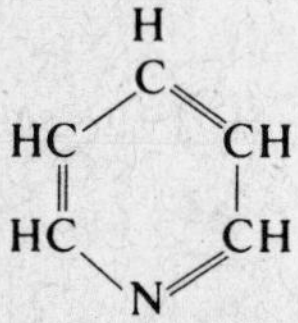

Pyridine 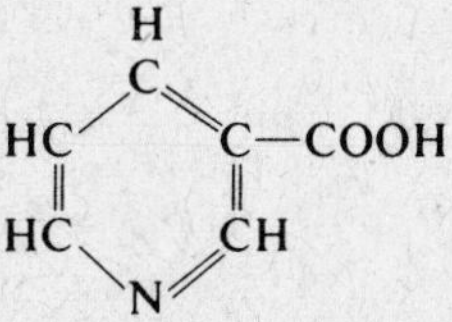　　　Nicotinic acid
(niacin)　　　Nicotinamide
(niacinamide)

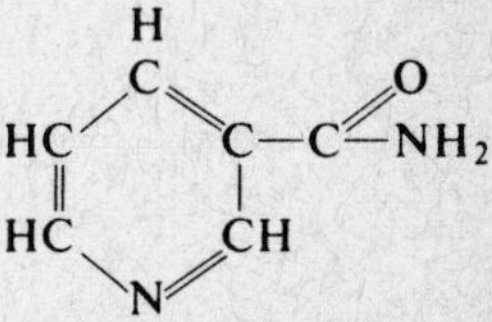

**Six-membered Rings with Two Hetero Atoms.**   *Pyrimidine* has two nitrogen atoms in the ring. Barbital is a derivative of pyrimidine. Thiamine (vitamin $B_1$) contains a pyrimidine group joined to a thiazole group. Derivatives of pyrimidine are important because they occur in nucleic acids.

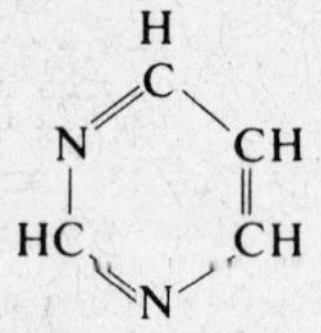

Pyrimidine
(1,3-diazine)

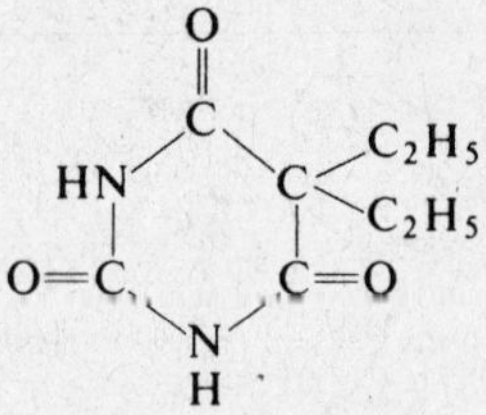

Barbital

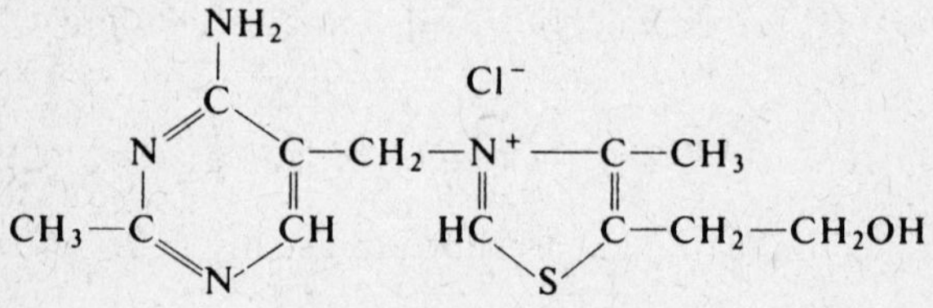

Thiamine hydrochloride
(vitamin $B_1$)

**Condensed Ring Systems.**   *Quinoline* is formed by the fusion of a benzene ring with pyridine. An important derivative of quinoline is the antimalarial drug quinine (see page 98). *Indole* is formed by the fusion of a benzene ring with a pyrrole ring. It is a product of intestinal putrefaction. Tryptophan, an amino acid, is a derivative of indole. *Purine* is a pyrimidine ring fused with an imidazole ring. The purine structure occurs in nucleic acids. Caffeine and uric acid are also derivatives of purine.

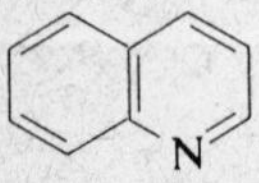

Quinoline

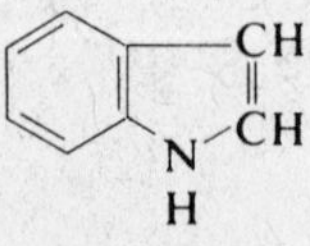

Indole

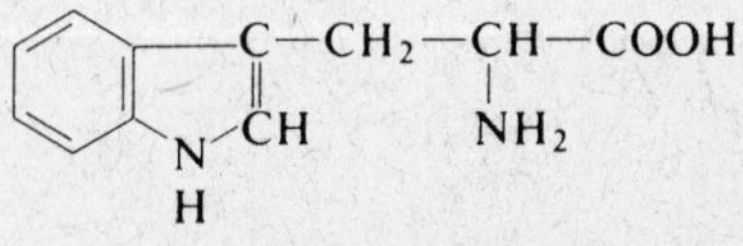

Tryptophan
(an amino acid)

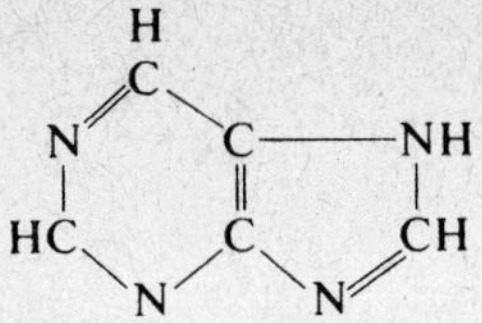

Purine

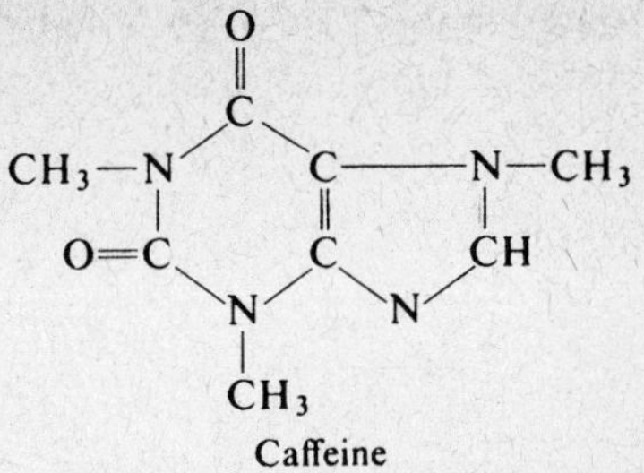

Caffeine

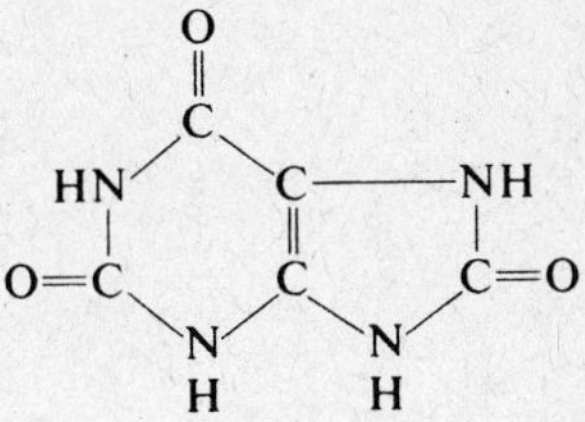

Uric acid

**Alkaloids.**    Alkaloids are basic, nitrogen-containing organic compounds found in plants.  They are usually heterocyclic, some having complicated ring structures based on several heterocyclic rings.  Most of the alkaloids are white crystalline solids with an intensely bitter taste.  They possess marked physiological activity and are valuable therapeutic agents.  Most alkaloids are insoluble in water but form soluble salts with acids; they are usually ad-

TABLE 16.1.    SOME COMMON ALKALOIDS

| Alkaloid | Source | Activity or Application |
|---|---|---|
| Atropine | Belladonna | Mydriatic, antispasmodic |
| Caffeine | Coffee | Stimulant, diuretic |
| Cinchonine | Cinchona bark | Antipyretic |
| Cocaine | Coca leaf | Local anesthetic |
| Codeine | Opium poppy | Cough control, analgesic |
| Ephedrine | *Ma huang* | Decongestant, mydriatic |
| Morphine | Opium poppy | Analgesic |
| Nicotine | Tobacco | Insecticide |
| Quinine | Cinchona bark | Antimalarial |
| Reserpine | Indian snake root (Rauwolfia) | Tranquilizer, sedative |
| Scopolamine | Roots of *Scopolia* | Sedative, hypnotic, mydriatic |
| Strychnine | Nux vomica | Poison, tonic |

ministered in the form of soluble hydrochlorides or sulfates. The structures of some common alkaloids are as follows:

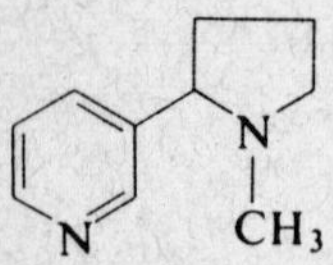

Nicotine

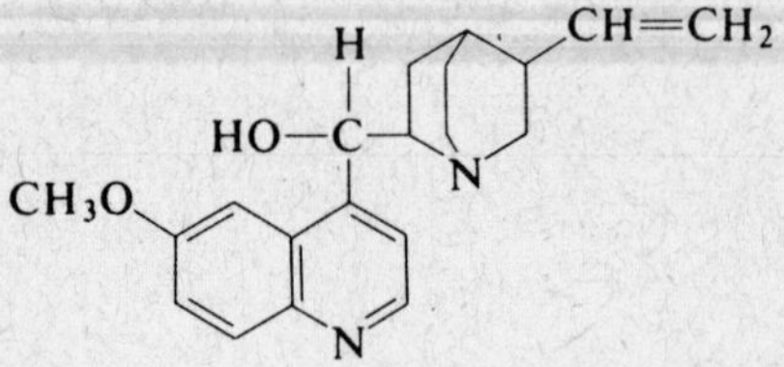

Quinine

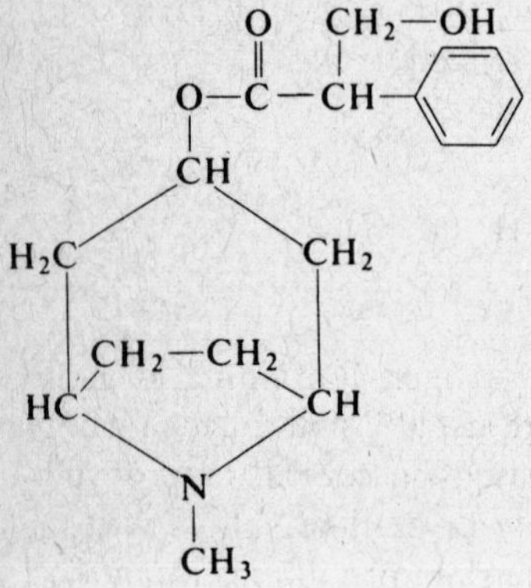

Atropine

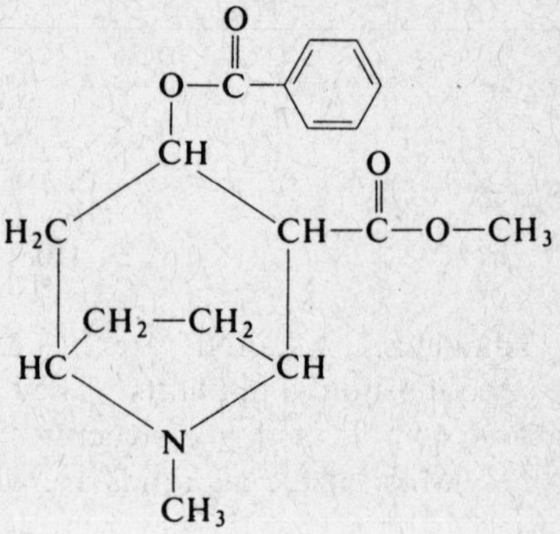

Cocaine

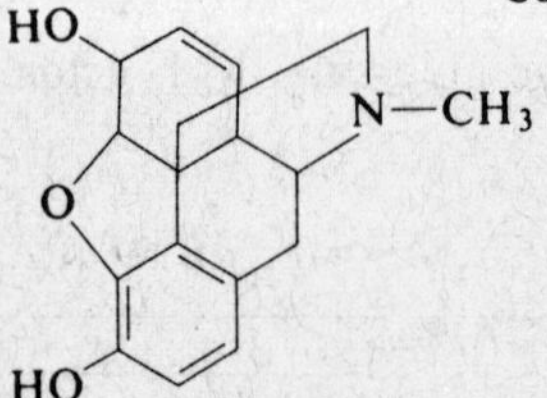

Morphine

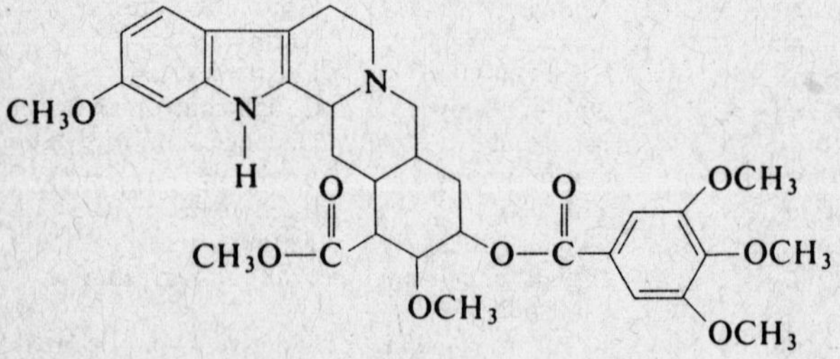

Reserpine

# 17. CARBOHYDRATES

The carbohydrates include the sugars, starches, cellulose, and other closely related substances. The class name "carbohydrates" designates these compounds as hydrates of carbon, since besides carbon, they contain hydrogen and oxygen in the ratio of two to one, as in water. However, there are some sugars that do not fit this general formula, and some compounds, such as formaldehyde ($CH_2O$), that fit the definition but are not carbohydrates. Carbohydrates are better defined as polyhydroxy aldehydes, or polyhydroxy ketones, or compounds that yield these substances on hydrolysis.

**Photosynthesis.** Carbohydrates are formed in the cells of plants, from carbon dioxide in the air and water in the ground. In the presence of sunlight and chlorophyll, the magnesium-containing green pigment of leaves, these two compounds react to form carbohydrate (represented by $C_6H_{12}O_6$ in the equation) and oxygen

$$6CO_2 + 6H_2O \xrightarrow[\text{chlorophyll}]{\text{sunlight}} C_6H_{12}O_6 + 6O_2$$

This process by which plants transform the radiant energy of the sun into chemical energy stored in food material is called *photosynthesis*. It is actually a series of complicated reactions. Photosynthesis is regarded by some as the most important chemical reaction in the world because it returns oxygen to the air as well as manufacturing and storing food material.

**Classification of Carbohydrates.** Carbohydrates are classified as monosaccharides, disaccharides, and polysaccharides on the basis of their hydrolysis possibilities.

*Monosaccharides* are carbohydrates that cannot be hydrolyzed. They may be further classified according to the number of carbon atoms in the molecule. Those with five carbon atoms are called pentoses ($C_5H_{10}O_5$); those with six, hexoses ($C_6H_{12}O_6$). The most important pentoses are arabinose, xylose, ribose, and deoxyribose, and the most important hexoses are glucose, galactose, and fructose.

*Disaccharides* ($C_{12}H_{22}O_{11}$) when hydrolyzed yield two molecules of monosaccharide for each molecule of disaccharide. The three important disaccharides are sucrose, maltose, and lactose.

*Polysaccharides* ($C_6H_{10}O_5)_x$ yield a large number of molecules of monosaccharide on complete hydrolysis. The polysaccharides include starch, glycogen, dextrin, cellulose, and pectins.

## MONOSACCHARIDES

The most important monosaccharides are glucose, galactose, and fructose. They can be represented by the following straight-chain formulas.

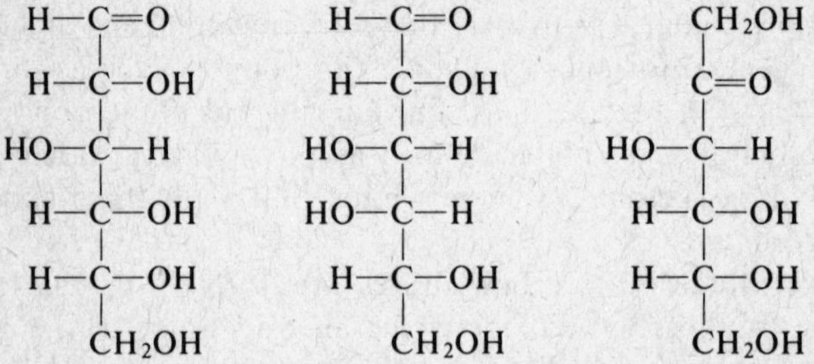

| Glucose | Galactose | Fructose |
|---------|-----------|----------|
| H—C=O | H—C=O | CH$_2$OH |
| H—C—OH | H—C—OH | C=O |
| HO—C—H | HO—C—H | HO—C—H |
| H—C—OH | HO—C—H | H—C—OH |
| H—C—OH | H—C—OH | H—C—OH |
| CH$_2$OH | CH$_2$OH | CH$_2$OH |

As shown above, glucose, galactose, and fructose have the same molecular formula ($C_6H_{12}O_6$) but different structural formulas. Glucose and galactose, whose molecules contain the aldehyde (H—C=O) group, are aldose sugars, while fructose, whose molecule contains the keto (C=O) group, is a ketose sugar.

These sugars exist principally in the cyclic form. The cyclic form of glucose, for example, is formed from the straight-chain form by the migration of the hydrogen of the —OH group on carbon 5 to the oxygen on carbon 1, and the joining of the oxygen on carbon 5 to carbon 1. In solution the cyclic form is in equilibrium with the straight-chain form.

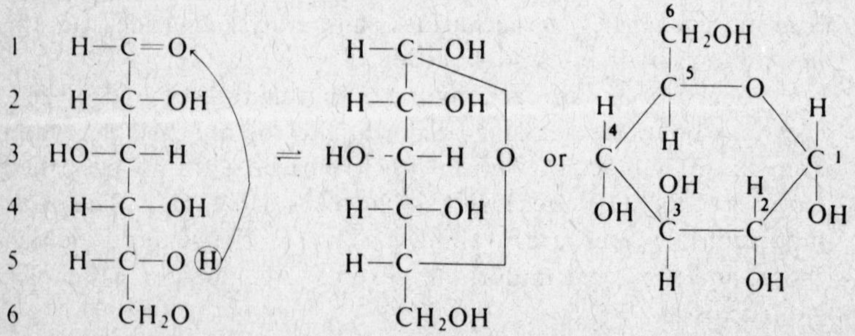

**Glucose.** Glucose, or *dextrose*, commonly known as *grape sugar*, is found in fruits and honey. It is the sugar of the blood and is found in the urine of diabetics. Glucose is prepared commercially by the acid hydrolysis of cornstarch. It is used for making candies, jellies, and other food products.

**Galactose.** Galactose does not occur in the free state in nature. It is found in combination in lactose, agar-agar, and pectin. It is less soluble and less sweet than glucose.

**Fructose.** Fructose, or *levulose*, is commonly known as *fruit sugar*. It occurs with glucose in fruits and honey. It is a constituent of sucrose and the polysaccharide inulin. It is the most soluble and the sweetest of all the sugars.

## DISACCHARIDES ($C_{12}H_{22}O_{11}$)

A disaccharide is formed when two molecules of monosaccharide join together with the loss of a molecule of water. The linkage is made from an aldehyde group of one monosaccharide to the ketone or hydroxyl group of the other. The most important disaccharides are sucrose, maltose, and lactose. Their structures are as follows:

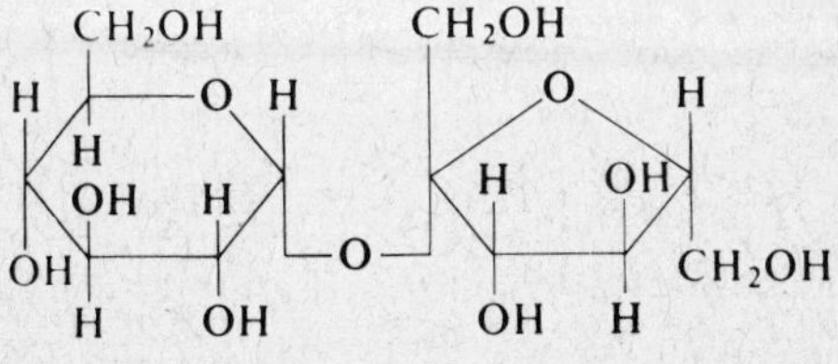

α-Glucose    α-Fructose

α-Sucrose

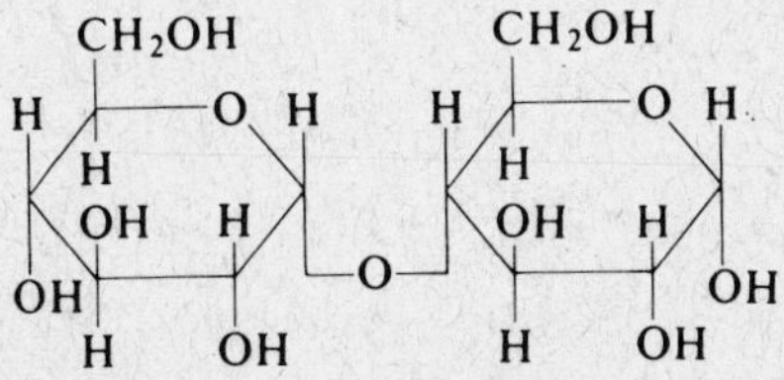

α-Glucose    α-Glucose

α-Maltose

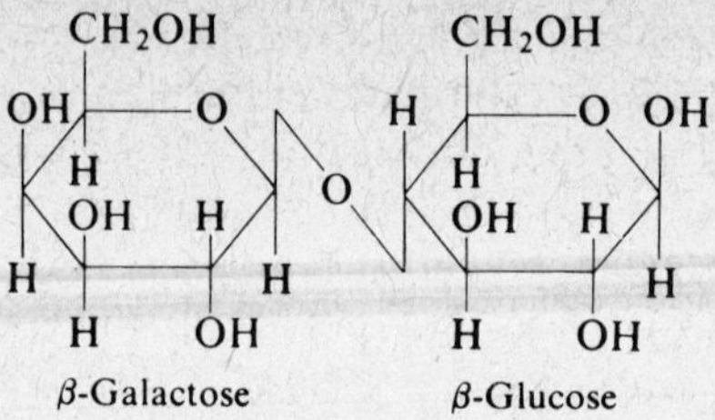

β-Galactose          β-Glucose

β-Lactose

**Sucrose.**   Sucrose is common table sugar.  It is prepared commercially from sugar cane or sugar beets.  It is also present in sorghum cane and the sap of sugar maple.  Sucrose is sweeter than glucose but not as sweet as fructose.  It is very soluble.  It is used in making candies and to sweeten food.

**Maltose.**   Maltose, or *malt sugar*, is present in germinating grains and in malt.  It is the product of the partial hydrolysis of starch by the enzyme diastase.  Maltose is less soluble and less sweet than sucrose.  It is used in feeding infants and invalids and in making malted milk and other malted products.

**Lactose.**   Lactose is also called *milk sugar* because it is present in milk.  Commercially it is obtained from milk whey, a by-product of the cheese industry.  It is the least soluble and least sweet of all common sugars.  The enzymes in certain bacteria will ferment lactose to lactic acid.  This occurs when milk sours.  Lactose is used in infant foods and special diets.

## POLYSACCHARIDES $(C_6H_{10}O_5)_x$

The polysaccharides are composed of many monosaccharide units joined together through the *glucoside linkage*.  A typical example is the amylose molecule present in starch, a portion of which is shown below.

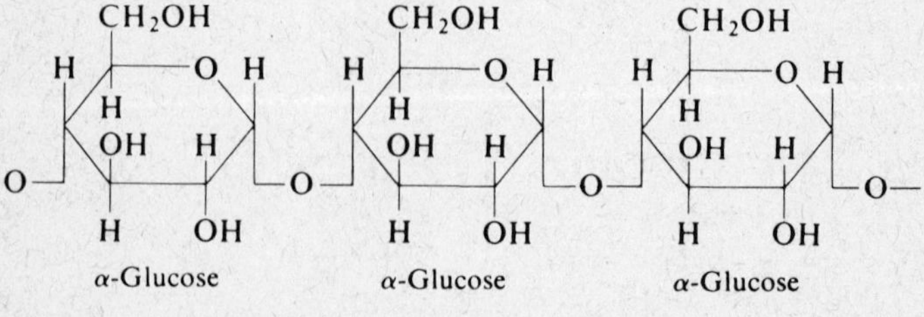

α-Glucose          α-Glucose          α-Glucose

**Starch.** Starch is the reserve food of plants, stored in the form of granules. These granules are insoluble in water but when ruptured by heating or grinding, they will form a colloidal dispersion. Starch gives a characteristic blue color with iodine. Chemically, the starch granule is composed of two types of molecules, amylose, a straight-chain polymer of $\alpha$-glucose, and amylopectin, a branched-chain polymer of $\alpha$-glucose. Starch, besides being the most important food polysaccharide, is used in laundering and in the manufacture of paste, sizing, alcohol, glucose, and corn syrup.

**Glycogen.** Glycogen, or *animal starch*, is the reserve food of animals. It is found in the liver and muscles and is especially abundant in shellfish. Unlike starch, glycogen is soluble in water and gives a red color with iodine. Chemically, it resembles amylopectin, but it is somewhat more highly branched.

**Dextrin.** Dextrin is a product of the partial hydrolysis of starch. It is manufactured by heating starch or treating it with an acid. It is more soluble than starch and has a slightly sweet taste. Dextrin is used in the confectionery industry and in making cosmetics, paints, varnish, sizing for textiles, and adhesives such as the mucilage for postage stamps.

**Cellulose.** Cellulose is the main constituent of the cell membranes of plants; therefore wood pulp, cotton, linen, straw, and hemp are the chief sources. Chemically, cellulose is a straight-chain polymer of $\beta$-glucose. It is used for the manufacture of paper, rayon, cellophane, cellulose acetate, nitrocellulose, ethyl cellulose, and similar products.

**Pectins.** Pectins are widely distributed in fruits and berries, notably in apples, grapes, and lemons. They yield galactose, arabinose, and galacturonic acid on hydrolysis. Fruit jellies are made by the proper balance of concentrations of pectins, sucrose, and fruit acid.

**Physical Properties of Carbohydrates.** The monosaccharides and disaccharides are white crystalline solids, sweet in taste, and soluble in water. They are *optically active*, that is, they rotate the plane of polarized light. Those that rotate the plane of polarized light to the right are said to be *dextrorotatory*, while those that rotate it to the left are called *levorotatory*. Glucose was formerly called dextrose, because it turns the plane of polarized light to the right ($+ 52°$). The old name for fructose was levulose, since it turns the plane of polarized light to the left ($-92°$). The presence of asymmetric carbon atoms in the molecule is responsible for the optical activity. An asymmetric carbon atom is one that is attached to four different groups.

The polysaccharides are amorphous and have little or no flavor. Because of their high molecular weights (50,000 to 5,000,000), they either form colloidal dispersions (starch) or are insoluble (cellulose).

**Chemical Properties of Carbohydrates.** The most important chemical properties of carbohydrates are hydrolysis, reducing action, and fermentation.

HYDROLYSIS. Disaccharides are hydrolyzed to two molecules of monosaccharides.

$$C_{12}H_{22}O_{11} + H_2O \rightarrow C_6H_{12}O_6 + C_6H_{12}O_6$$

Sucrose     $\rightarrow$ Glucose + Fructose

Lactose     $\rightarrow$ Glucose + Galactose

Maltose     $\rightarrow$ Glucose + Glucose

Common polysaccharides when completely hydrolyzed yield glucose. Stages in the hydrolysis of starch may be followed by observing the color of the compounds formed with iodine, which is blue for starch and amylodextrin, red for erythrodextrin, and colorless for achrodextrin, maltose, and glucose.

$$(C_6H_{10}O_5)_n \xrightarrow{+ H_2O} (C_6H_{10}O_5)_x \xrightarrow{+ H_2O} (C_6H_{10}O_5)_y \xrightarrow{+ H_2O}$$
Starch            Amylodextrin        Erythrodextrin

$$(C_6H_{10}O_5)_z \xrightarrow{+ H_2O} C_{12}H_{22}O_{11} \xrightarrow{+ H_2O} C_6H_{12}O_6$$
Achrodextrin          Maltose              Glucose

In the above equations, the subscript $n$ stands for a larger number than $x$, $x$ for a larger number than $y$, and $y$ for a larger number than $z$, indicating that the molecule gets smaller and smaller as hydrolysis progresses.

REDUCING ACTION. Sugars that contain aldehyde or keto groups (e.g., glucose, fructose, galactose, maltose, and lactose) are called *reducing sugars*. They reduce the soluble blue cupric ion in Fehling's or Benedict's solution to the insoluble orange-red cuprous oxide, they themselves being oxidized to acids. This is the basis of the clinical test for sugar in the urine.

$$\underset{\text{Glucose}}{\overset{\displaystyle \text{CHO}}{\underset{\displaystyle \text{CH}_2\text{OH}}{(\text{H}-\text{C}-\text{OH})_4}}} + \underset{\text{Reagent}}{2\,\text{Cu(OH)}_2} \rightarrow \underset{\substack{\text{Cuprous} \\ \text{oxide}}}{\text{Cu}_2\text{O}\downarrow} + \underset{\text{Gluconic acid}}{\overset{\displaystyle \text{COOH}}{\underset{\displaystyle \text{CH}_2\text{OH}}{(\text{H}-\text{C}-\text{OH})_4}}} + 2\,\text{H}_2\text{O}$$

Sucrose is a *nonreducing sugar.* The glucose and fructose molecules that make up the sucrose molecule are joined to each other in such a way that both the aldehyde and keto groups present in the original sugars have been eliminated.

FERMENTATION.   Simple sugars (with the exception of galactose and lactose) are fermented by the enzyme *zymase*, present in yeast, to alcohol and carbon dioxide.   Disaccharides (with the exception of lactose) and polysaccharides have to be hydrolyzed to monosaccharides before they can be fermented.

$$\underset{\text{Glucose}}{C_6H_{12}O_6} \xrightarrow{\text{zymase}} \underset{\text{Alcohol}}{2\,C_2H_5OH} + \underset{\substack{\text{Carbon} \\ \text{dioxide}}}{2\,CO_2}$$

# *18.* LIPIDS

Lipids, also called lipides, lipins, or lipoids, are a large group of natural fatlike substances that are insoluble in water but soluble in organic solvents such as ether, chloroform, and benzene.

**Classification.** Lipids can be divided into three main groups. *Simple lipids* are esters of fatty acids and alcohols, which they yield on hydrolysis. Fats and waxes are members of this group. *Compound lipids* yield on hydrolysis some other product besides alcohols and fatty acids. Phospholipids and glycolipids are the most important lipids of this type. *Derived lipids* are obtained by the hydrolysis of simple or compound lipids. Sterols and the fatty acids themselves are derived lipids.

## FATTY ACIDS

Fatty acids are usually straight-chain carboxylic acids which can be saturated or unsaturated. Those found in nature almost always have an even number of carbon atoms in their molecules.

### TABLE 18.1. COMMON FATTY ACIDS

| Name | Formula | Some Sources |
|------|---------|--------------|
| SATURATED: | | |
| Butyric | $C_3H_7COOH$ | Butter |
| Caproic | $C_5H_{11}COOH$ | Butter, coconut oil |
| Caprylic | $C_7H_{15}COOH$ | Butter, coconut oil |
| Capric | $C_9H_{19}COOH$ | Butter, coconut oil |
| Lauric | $C_{11}H_{23}COOH$ | Coconut oil |
| Myristic | $C_{13}H_{27}COOH$ | Coconut oil, nutmeg butter |
| Palmitic | $C_{15}H_{31}COOH$ | Animal and vegetable fats |
| Stearic | $C_{17}H_{35}COOH$ | Animal and vegetable fats |
| UNSATURATED: | | |
| Oleic (1 C=C)* | $C_{17}H_{33}COOH$ | Olive oil |
| Linoleic (2 C=C) | $C_{17}H_{31}COOH$ | Corn oil, soybean oil |
| Linolenic (3 C=C) | $C_{17}H_{29}COOH$ | Linseed oil |
| Arachidonic (4 C=C) | $C_{19}H_{31}COOH$ | Liver |

*Number of double bonds. Oleic acid containing one double bond is *monounsaturated*. Linoleic, linolenic, and arachidonic acids containing more than one double bond are *polyunsaturated*.

Palmitic and stearic acids, which are found in animal and vegetable fats, are the most important saturated fatty acids. Oleic acid, which contains one double bond, is the most common unsaturated fatty acid. Linoleic, linolenic, and arachidonic acids, which have two, three, and four double bonds, respectively, are of special nutritional importance and are known as *essential fatty acids*. The principal fatty acids that occur in nature are listed in Table 18.1.

## FATS

Fats are esters of the trihydric alcohol glycerol and fatty acids. If all the fatty acids in the molecule are of the same kind, the fat is called a *simple glyceride*; if they are different, the fat is a *mixed glyceride*.

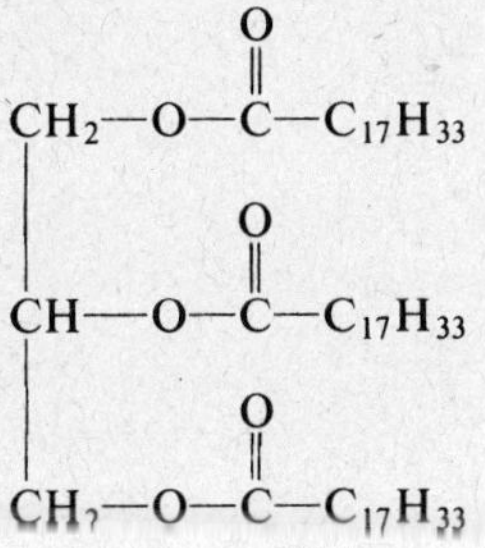

Glyceryl trioleate
or Triolein
(a simple glyceride)

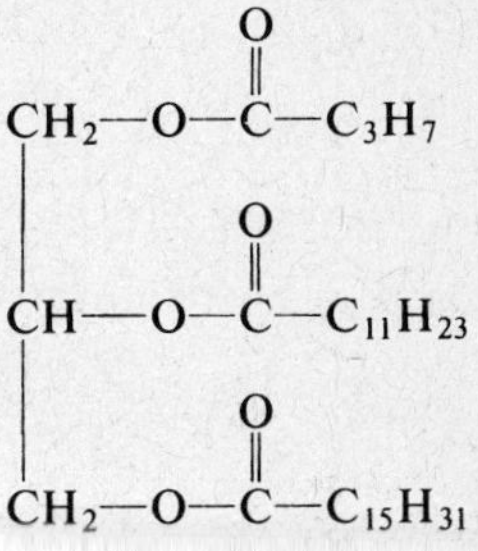

Glyceryl butyro-
lauro-palmitate
(a mixed glyceride)

The formation of fat from glycerol and fatty acids may be represented by the following equation, in which the R's represent the same or different hydrocarbon radicals:

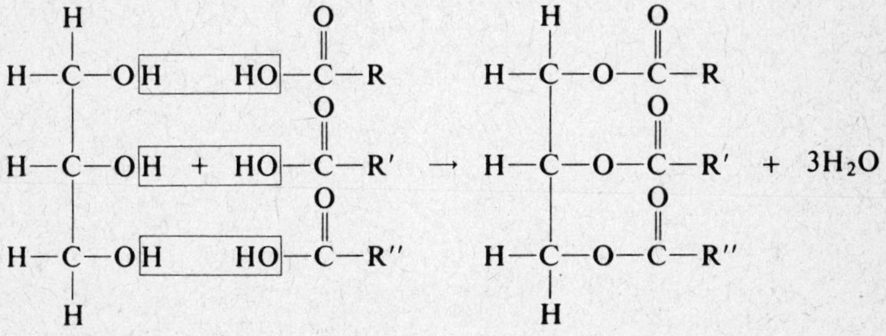

Glycerol        Fatty acids                 A fat

**Physical Properties of Fats.** Fats composed largely of the glycerides of the higher saturated fatty acids are solids at room temperature. Most of the animal fats are of this type. Fats derived largely from unsaturated fatty acids, such as the vegetable oils, are liquids at room temperature.

Fats have a characteristic greasy feeling and form a transparent grease spot when placed on a sheet of paper (*spot test*). They are lighter than water and insoluble in it, but are soluble in fat solvents such as hot alcohol, ether, chloroform, and benzene.

**Chemical Properties of Fats.** Because of their ester linkages, fats can be hydrolyzed and saponified. Unsaturated fats will form addition products with oxygen, hydrogen, or halogens.

HYDROLYSIS. When hydrolyzed, fats yield glycerol and fatty acids.

$$
\begin{array}{l}
CH_2-O-\overset{\displaystyle O}{\overset{\|}{C}}-C_{17}H_{35} \\[2mm]
CH-O-\overset{\displaystyle O}{\overset{\|}{C}}-C_{17}H_{35} + 3H_2O \rightarrow \\[2mm]
CH_2-O-\overset{\displaystyle O}{\overset{\|}{C}}-C_{17}H_{35}
\end{array}
\qquad
\begin{array}{l}
CH_2OH \\[2mm]
CHOH + 3C_{17}H_{35}COOH \\[2mm]
CH_2OH
\end{array}
$$

Tristearin          Glycerol          Stearic acid

SAPONIFICATION. Upon saponification (alkaline hydrolysis), fats yield glycerol and soap (the alkali salt of high fatty acids).

$$
\begin{array}{l}
CH_2-O-\overset{\displaystyle O}{\overset{\|}{C}}-C_{15}H_{31} \\[2mm]
CH-O-\overset{\displaystyle O}{\overset{\|}{C}}-C_{15}H_{31} + 3NaOH \rightarrow \\[2mm]
CH_2-O-\overset{\displaystyle O}{\overset{\|}{C}}-C_{15}H_{31}
\end{array}
\qquad
\begin{array}{l}
CH_2OH \\[2mm]
CHOH + 3C_{15}H_{31}COONa \\[2mm]
CH_2OH
\end{array}
$$

Tripalmitin          Glycerol          Sodium palmitate
                                        (a soap)

The *saponification number* of a fat is the milligrams of potassium hydroxide required to saponify one gram of fat. It is used to determine the size of the molecule: the smaller the molecule, the higher the saponification number.

HYDROGENATION. Liquid vegetable oils may be transformed into solid fats by saturating the unsaturated acid radicals in the oil with hydrogen in the presence of a nickel catalyst at about 200°C. Two hydrogen atoms are added to each double bond. Solid shortenings like Crisco and Spry are made from vegetable oils by this process.

ADDITION OF IODINE. Iodine, like hydrogen, can add to the double bond that is present in the unsaturated fat. The degree of unsaturation of a fat or an oil can be determined by finding its *iodine number*, which is defined as the number of grams of iodine absorbed by 100 grams of fat or oil.

OXIDATION. Fats may become rancid when they oxidize slowly in air; the short-chained aldehydes and ketones that are formed have unpleasant odors. This type of rancidity is called *oxidative rancidity*, as distinguished from *hydrolytic rancidity* which occurs when saturated fats are hydrolyzed to form fatty acids with an objectionable taste (e.g., butyric acid in rancid butter).

Air oxidation accounts for the drying property of the highly unsaturated oils, such as linseed oil and tung oil, when they are used in paints and varnishes.

FORMATION OF ACROLEIN. When a fat is heated to a very high temperature, it decomposes. The glycerol that is liberated is dehydrated to acrolein, an unsaturated aldehyde with a sharp, penetrating odor characteristic of burned fat. In the laboratory, this *acrolein test* is carried out by heating the fat with a dehydrating agent such as potassium bisulfate.

$$\begin{array}{ccc} CH_2OH & & H-C=O \\ | & \xrightarrow{KHSO_4} & \\ CHOH & & CH \quad + 2H_2O \\ | & & \| \\ CH_2OH & & CH_2 \\ \text{Glycerol} & & \text{Acrolein} \end{array}$$

## SOAPS AND DETERGENTS

Soaps are metallic salts of high fatty acids. Ordinary *hard soaps* are sodium salts of high fatty acids or sodium soaps, while *soft soaps* are potassium soaps. *Castile soap* is pure sodium oleate, from olive oil. *Tincture of green soap* is a solution of a soft soap in alcohol.

The cleansing property of soaps is due to their ability to lower the surface tension of water and act as emulsifying agents; grease on the skin or fabric is emulsified to small particles and carried away by the wash water. Any dirt clinging to the grease is likewise washed away. Soaps also have antiseptic properties.

While sodium or potassium soaps are soluble in water, calcium or magnesium soaps are insoluble. Hence hard water containing calcium or magnesium ions will precipitate out soluble sodium or potassium soap to form scums.

Detergents are synthetic products used as soap substitutes. They do not form precipitates in hard water. Two of the most common types of detergents are the sodium alkyl sulfates (e.g., sodium lauryl sulfate, $C_{12}H_{25}OSO_3Na$) and the alkyl aryl sulfonates (e.g., sodium undecyl benzene sulfonate, $C_{11}H_{23}C_6H_5$-$SO_3Na$).

## WAXES

Waxes are esters of long-chain fatty acids and high molecular weight alcohols. *Beeswax*, from the honeycomb, and *lanolin*, from wool, are widely used as a base for ointments, salves, and creams. *Spermaceti*, secreted by the sperm whale, is used in the manufacture of candles, cosmetics, and pharmaceutical products. *Carnauba wax*, obtained from the carnauba palm, is used in floor waxes and in automobile and furniture polishes.

## PHOSPHOLIPIDS

Phospholipids, or phosphatides, are complex esters composed of an alcohol, fatty acids, phosphoric acid, and a nitrogenous base. They are present in every tissue of the body, but especially in the nervous system. The three main types of phospholipids are the lecithins, cephalins, and sphingomyelins.

**Lecithins.** Lecithins resemble fats, but one of the three fatty acids in the molecule is replaced by phosphoric acid joined to the nitrogenous base choline ($HO-CH_2-CH_2-N(CH_3)_3-OH$). Lecithins are found in liver, brain and nerve tissue, egg yolk, and soybeans. They are good emulsifying agents and aid in the absorption and transport of fatty acids in the body, as well as serving as a source of phosphoric acid for the synthesis of cells. They are used in the manufacture of candy, margarine, aviation gasoline, and medicines.

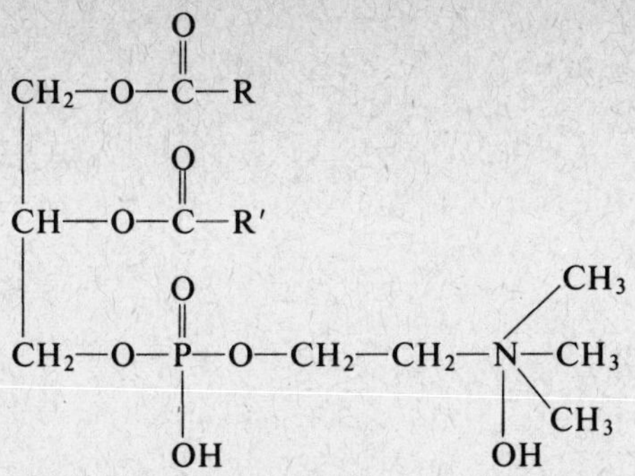

$$CH_2-O-\overset{\displaystyle O}{\overset{\displaystyle \|}{C}}-R$$

$$CH-O-\overset{\displaystyle O}{\overset{\displaystyle \|}{C}}-R'$$

$$CH_2-O-\overset{\displaystyle O}{\overset{\displaystyle \|}{P}}-O-CH_2-CH_2-\overset{\displaystyle CH_3}{\underset{\displaystyle OH}{N}}\overset{\displaystyle -CH_3}{\underset{\displaystyle CH_3}{}}$$
$$\underset{\displaystyle OH}{}$$

A Lecithin

**Cephalins.** Cephalins resemble lecithins except that the choline is replaced by ethanolamine ($HO-CH_2-CH_2-NH_2$) in phosphatidylethanolamine, by serine ($HOCH_2CHNH_2COOH$) in phosphatidylserine, and by inositol (1,2,3,4,5,6-hexahydroxycyclohexane) in phosphoinositide. Cephalins are found in liver, brain, blood, and yeast. Some cephalins are essential for the coagulation of blood. They also serve as a source of phosphoric acid.

**Sphingomyelins.** Sphingomyelins resemble lecithins except that the glycerol is replaced by the amino alcohol sphingosine, ($CH_3-(CH_2)_{12}-CH=CH-CHOH-CHNH_2-CH_2OH$), and there is only one fatty acid molecule. Sphingomyelins are found in brain, nervous tissue, lungs, and spleen.

## GLYCOLIPIDS

Glycolipids, or cerebrosides, are complex lipids containing sphingosine, a fatty acid, and a sugar (galactose or glucose), but no phosphoric acid. They are found in the brain, nervous tissue, kidneys, lungs, spleen, liver, egg yolk, and fish sperm.

## STEROLS

Sterols are high molecular weight cyclic alcohols. They exist in the solid state. They are important constituents of plant and animal tissues.

*Cholesterol* is the most important animal sterol and has the structure:

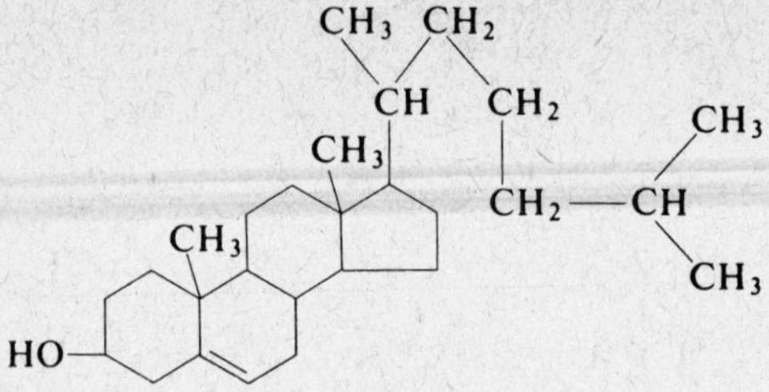

Cholesterol

Cholesterol is found in the brain, nervous tissue, bile, gallstones, egg yolk, butter, and animal fats. Excessive intake of cholesterol-rich foods or saturated fats is believed by some to cause atherosclerosis and coronary heart disease, but the evidence is inconclusive.

*Ergosterol*, an important plant sterol, has two more double bonds and one more methyl group than cholesterol. It is found in yeast, certain mushrooms, and ergot, a fungus of rye. It is converted to vitamin $D_2$ when irradiated by ultraviolet light.

Other compounds containing the sterol structure are the sex hormones, the adrenal cortex hormones, and the bile acids or salts.

# 19. PROTEINS

Proteins are complex nitrogenous substances of high molecular weight. The word protein is derived from the Greek word *proteios* meaning "of first importance." The proteins are the fundamental constituents of living cells and are essential to the growth and repair of all tissues. Plants can synthesize proteins from simple inorganic substances found in the air and soil, but animals depend largely on organic sources for their protein needs.

The elements present in nearly all proteins are carbon, hydrogen, oxygen, nitrogen, and sulfur. Other elements may also be present in specific proteins, e.g., iron in hemoglobin, iodine in thyroglobulin, and phosphorus in casein.

Protein molecules are very large, their molecular weights ranging from 10,000 to 25,000,000. For example, oxyhemoglobin, whose molecular formula is $(C_{783}H_{1166}O_{208}N_{203}S_2Fe)_4$, has a molecular weight of 65,000.

**Amino Acids.** The building blocks that make up the large protein molecule are the amino acids. Amino acids are organic acids that contain an amino group, $—NH_2$, as well as a carboxyl group, $—COOH$. In all the naturally occurring amino acids the amino group is attached to the $\alpha$-carbon atom, which is the carbon atom next to the carboxyl group.

Over twenty different amino acids are known. The simplest amino acid is glycine, which has the following formula.

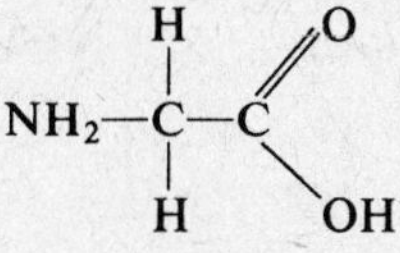

Glycine

The general formula for an amino acid is

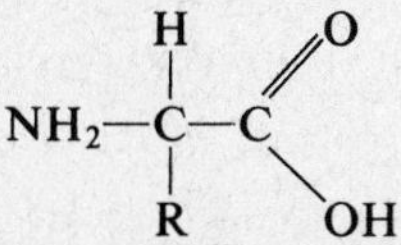

113

where side chains (R) of varying complexity are in the place of one of the hydrogen atoms in glycine.

Amino acids can be aliphatic, aromatic, or heterocyclic. They may have one or two amino groups and one or two carboxyl groups. Several have hydroxy groups and sulfur atoms. The formulas of some amino acids are given in Table 19.1.

### TABLE 19.1.   CLASSIFICATION AND STRUCTURE OF SOME COMMON AMINO ACIDS

#### ALIPHATIC MONOAMINO MONOCARBOXYLIC ACIDS

Glycine

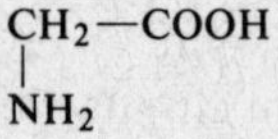

$$CH_2-COOH$$
$$|$$
$$NH_2$$

Alanine

$$CH_3-CH-COOH$$
$$|$$
$$NH_2$$

Valine

$$CH_3-CH-CH-COOH$$
$$|\quad\ |$$
$$CH_3\ NH_2$$

Leucine

$$CH_3-CH-CH_2-CH-COOH$$
$$|\qquad\qquad |$$
$$CH_3\qquad\quad NH_2$$

Isoleucine

$$CH_3-CH_2-CH-CH-COOH$$
$$|\quad\ |$$
$$CH_3\ NH_2$$

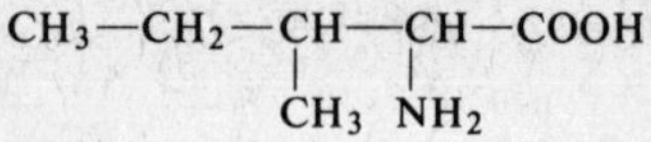

#### HYDROXY AMINO ACIDS

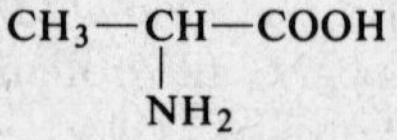

Serine

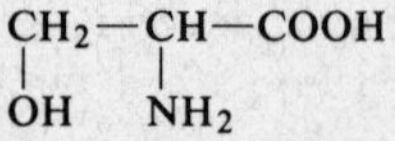

$$CH_2-CH-COOH$$
$$|\quad\ |$$
$$OH\quad NH_2$$

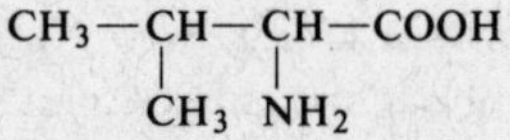

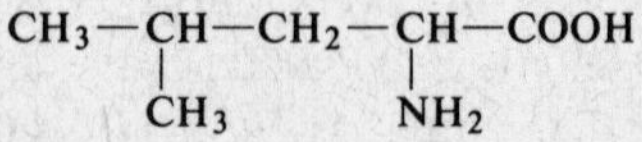

Threonine

$$CH_3-CH-CH-COOH$$
$$|\quad\ |$$
$$OH\quad NH_2$$

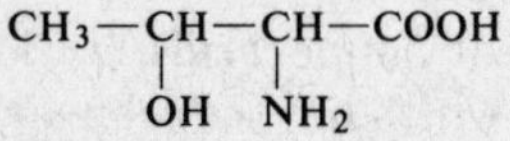

#### ALIPHATIC MONOAMINO DICARBOXYLIC ACIDS

Aspartic acid

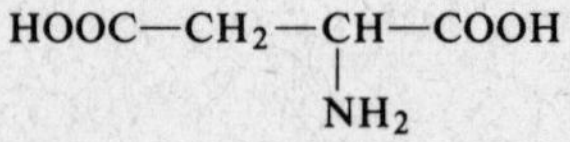

$$HOOC-CH_2-CH-COOH$$
$$|$$
$$NH_2$$

Glutamic acid

$$HOOC-CH_2-CH_2-CH-COOH$$
$$|$$
$$NH_2$$

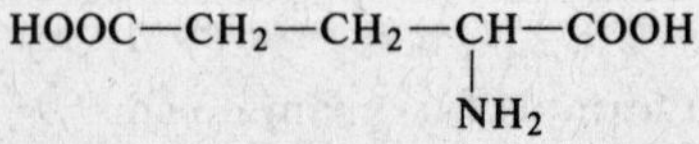

#### ALIPHATIC DIAMINO MONOCARBOXYLIC ACIDS

Lysine

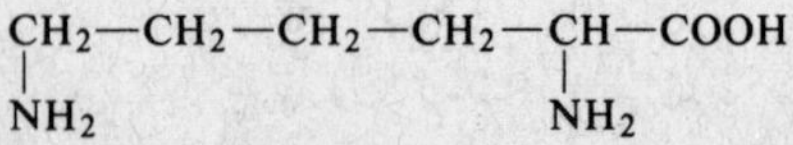

$$CH_2-CH_2-CH_2-CH_2-CH-COOH$$
$$|\qquad\qquad\qquad\qquad\quad |$$
$$NH_2\qquad\qquad\qquad\qquad NH_2$$

## TABLE 19.1.   CLASSIFICATION AND STRUCTURE OF SOME COMMON AMINO ACIDS (Continued)

Arginine

$$H_2N-C-NH-CH_2-CH_2-CH_2-CH-COOH$$
$$\quad\quad |\quad\quad\quad\quad\quad\quad\quad\quad\quad\quad\quad |$$
$$\quad\quad NH\quad\quad\quad\quad\quad\quad\quad\quad\quad NH_2$$

### SULFUR AMINO ACIDS

Cysteine

$$CH_2-SH$$
$$|$$
$$H-C-NH_2$$
$$|$$
$$COOH$$

Cystine

$$CH_2-S-S-CH_2$$
$$|\quad\quad\quad\quad\quad\quad |$$
$$H-C-NH_2\quad H-C-NH_2$$
$$|\quad\quad\quad\quad\quad\quad |$$
$$COOH\quad\quad\quad COOH$$

Methionine

$$CH_3-S-CH_2-CH_2-CH-COOH$$
$$\quad\quad\quad\quad\quad\quad\quad\quad\quad\quad |$$
$$\quad\quad\quad\quad\quad\quad\quad\quad\quad NH_2$$

### AROMATIC AMINO ACIDS

Phenylalanine

$$\bigcirc CH_2-CH-COOH$$
$$\quad\quad\quad\quad |$$
$$\quad\quad\quad NH_2$$

Tyrosine

$$HO\bigcirc CH_2-CH-COOH$$
$$\quad\quad\quad\quad\quad\quad |$$
$$\quad\quad\quad\quad\quad NH_2$$

### HETEROCYCLIC AMINO ACIDS

Proline

$$CH_2 \quad\quad\quad CH_2$$
$$|\quad\quad\quad\quad\quad\quad |$$
$$CH_2 \quad\quad\quad CH-COOH$$
$$\quad\backslash \quad N \quad /$$
$$\quad\quad |$$
$$\quad\quad H$$

Hydroxyproline

$$HO-CH \quad\quad\quad CH_2$$
$$|\quad\quad\quad\quad\quad\quad\quad |$$
$$CH_2 \quad\quad\quad CH-COOH$$
$$\quad\backslash \quad N \quad /$$
$$\quad\quad |$$
$$\quad\quad H$$

Histidine

$$HC=C-CH_2-CH-COOH$$
$$|\quad\quad |\quad\quad\quad\quad\quad |$$
$$N\quad\quad NH\quad\quad\quad\quad NH_2$$
$$\quad\backslash C /$$
$$\quad\quad |$$
$$\quad\quad H$$

Tryptophan

$$\bigcirc\bigcirc CH_2-CH-COOH$$
$$\quad | \quad\quad\quad\quad\quad\quad |$$
$$\quad N \quad\quad\quad\quad\quad NH_2$$
$$\quad |$$
$$\quad H$$

*Essential amino acids* are those indispensable to human nutrition which cannot be synthesized in the body from simpler nitrogenous compounds, either at all or in sufficient quantity, and must therefore be supplied by the protein in the diet. They are lysine, tryptophan, phenylalanine, leucine, isoleucine, threonine, methionine, and valine. Arginine and histidine have been found to be essential in the rat.

**Structure of Protein.** The amino acids in the protein molecule are joined together through the amino group of one and the carboxyl group of the other, with the loss of one molecule of water. This is called the *peptide linkage*, $-\overset{\overset{\textstyle O}{\|}}{C}-\overset{\overset{\textstyle H}{|}}{N}-$. Below is the structure of a portion of a protein molecule showing the peptide linkage.

$$-\overset{\overset{\textstyle H}{|}}{N}-CH_2-\overset{\overset{\textstyle O}{\|}}{C}-\overset{\overset{\textstyle H}{|}}{N}-\underset{\underset{\textstyle CH_3}{|}}{CH}-\overset{\overset{\textstyle O}{\|}}{C}-\overset{\overset{\textstyle H}{|}}{N}-\underset{\underset{\textstyle CH_2}{|}}{CH}-\overset{\overset{\textstyle O}{\|}}{C}-$$

Glycine        Alanine        Tyrosine

When two amino acids are united by a peptide linkage, the compound is called a *dipeptide*; when three are joined, it is a *tripeptide*. A compound resulting from the linkage of many amino acids is called a *polypeptide*. Each protein is a unique sequence of amino acids joined by peptide bonds.

In unstretched fibrous proteins like wool and leather, the polypeptide chain is coiled in a spiral structure known as the *α-helix*. Each turn of the spiral is held together by *hydrogen bonding* between amino nitrogen and carbonyl oxygen. Another type of bonding is the *disulfide linkage* ($-S-S-$) found in cystine. In silk and in stretched hair, there is a sheetlike network of polypeptide chains lying side by side in a *pleated-sheet structure*. It is thought that when hair is stretched, some of the hydrogen bonds are broken and replaced by hydrogen bonds between adjacent chains, changing it from an α-helix structure to a pleated-sheet structure. The α-helix and pleated-sheet structures are shown in Figs. 19.1 and 19.2, respectively.

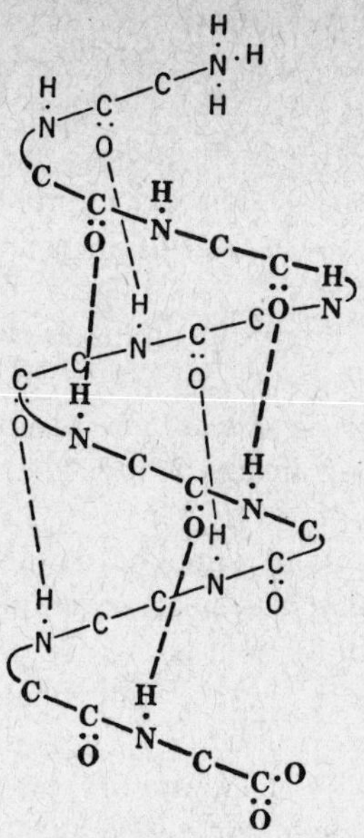

FIG. 19.1. The α-helix structure of protein.    Dash lines indicate hydrogen bonds.  From *Elements of General and Biological Chemistry* by John R. Holum. Copyright 1962, John Wiley & Sons, Inc.

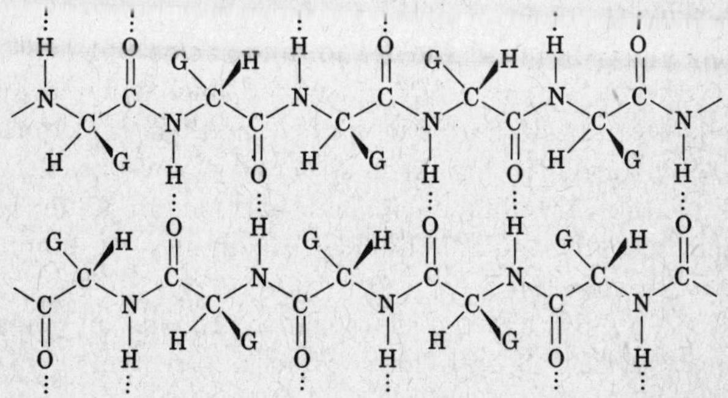

FIG. 19.2. The pleated-sheet structure of protein, showing hydrogen bonding between extended polypeptide chains.  From *Elements of General and Biological Chemistry* by John R. Holum.  Copyright 1962, John Wiley & Sons, Inc.

The *fibrous proteins*, such as those in hair and muscle, have a fairly stable, elongated structure and are insoluble in water.  In *globular proteins*, such as hemoglobin and albumin, the poly-peptide helix itself is looped and folded into a compact structure which forms a colloidal solution in water. The $\alpha$-helix structure is believed to be present in some globular proteins as well as in fibrous ones.

**Classification of Proteins.**    Proteins are usually classified into groups according to their chemical composition and solubility. There are three main divisions: simple proteins, conjugated proteins, and derived proteins.

SIMPLE PROTEINS.    Simple proteins are those that yield only amino acids on hydrolysis.

*Albumins* are soluble in water and coagulated by heat. Examples are egg albumin in egg white, lactalbumin in milk, and serum albumin in blood.

*Globulins* are insoluble in water but soluble in dilute salt solutions. They are coagulated by heat. Examples are serum globulin in blood and lactoglobulin in milk.

*Glutelins* are insoluble in water and neutral salt solutions but soluble in dilute alkalies and acids. They are coagulated by heat. Examples are glutenin in wheat and oryzenin in rice.

*Prolamines* are soluble in 70 to 80 per cent alcohol but insoluble in water or absolute alcohol. Examples are gliadin in wheat and zein in corn.

*Albuminoids* are insoluble in all the solvents listed above. Examples are keratin in hair, horn, feather and nail; elastin in tendons; collagen in bone.

*Histones* are soluble in water and dilute acids but insoluble in dilute ammonium hydroxide. They are uncoagulable by heat. Examples are the globin of hemoglobin and thymus histone.

*Protamines* are the simplest in structure of the proteins. They are soluble in water, dilute acids, and dilute ammonium hydroxide. They are uncoagulable by heat. Examples are salmine and sturine in the sperm of fish.

CONJUGATED PROTEINS.    Conjugated proteins are those that on hydrolysis yield nonprotein substances (prosthetic groups) in addition to amino acids.

*Nucleoproteins* are protein linked with nucleic acid. They are found in the nuclei of all living cells.

*Glycoproteins* are protein linked with a carbohydrate. Examples are mucin in saliva and mucoids in tendon and cartilage.

*Phosphoproteins* are protein linked with phosphoric acid. Examples are casein in milk and vitellin in egg yolk.

*Chromoproteins* are protein linked with a color-producing group. Examples are hemoglobin in blood and melanin in hair and feathers.

*Lipoproteins* are protein linked with lecithin or cholesterol. Examples are fibrinogen and a protein in egg yolk.

DERIVED PROTEINS. Derived proteins are substances produced by the action of heat, acids, alkalies, or enzymes on simple and conjugated proteins. This group includes all the hydrolysis products of the original protein.

*Proteans* are the first hydrolysis products of the action of dilute acids or enzymes.

*Metaproteins* are further hydrolysis products formed by the action of dilute acids or alkalies. They are soluble in weak acids and alkalies. Examples are acid metaprotein (acid albuminate) and alkali metaprotein (alkali albuminate).

*Coagulated proteins* are insoluble products formed by the action of heat or alcohol. An example is cooked egg albumin.

*Proteoses, peptones,* and *peptides* are products of the further hydrolysis of proteins, listed in the order of decreasing complexity. The proteoses are soluble in water, uncoagulable by heat, and precipitated in a saturated ammonium sulfate solution. The peptones have properties similar to those of the proteoses, but they are not precipitated in an ammonium sulfate solution. The peptides are combinations of two or more amino acids and are thus the simplest derived protein molecules.

**Physical Properties of Proteins.** Proteins form colloidal solutions and do not pass through animal membranes such as the intestine and the glomerulus of the kidney, unless the membranes are damaged. Proteins are hydrophilic colloids and have the ability to adsorb large quantities of water.

**Chemical Properties of Proteins.** Because the amino acids have a great variety of chemically reactive groups, and because some of the cross-linkages holding the protein molecule together are relatively weak, proteins have a wide range of chemical activity and are sensitive to many chemical reagents and environmental conditions.

AMPHOTERIC PROPERTIES AND ISOELECTRIC POINT. Amino acids, and therefore proteins, are *amphoteric*, which means that they can act both as proton donors and as proton acceptors. This is because their amino group is basic and their carboxyl group is acidic. Amino acids are believed to exist in solution in the form of dipolar ions or *zwitterions*, which can neutralize both acids and bases as shown below.

$$CH_3\!-\!CH\!-\!COO^- + H^+Cl^- \rightarrow CH_3\!-\!CH\!-\!COOH + Cl^-$$
$$\quad\quad\;\; NH_3^+ \quad\quad\quad\quad\quad\quad\quad\quad\quad\;\; NH_3^+$$

Zwitterion          An acid

$$CH_3\!-\!CH\!-\!COO^- + Na^+OH^- \rightarrow CH_3\!-\!CH\!-\!COO^-Na^+ + H_2O$$
$$\quad\quad\;\; NH_3^+ \quad\quad\quad\quad\quad\quad\quad\quad\quad\;\; NH_2$$

Zwitterion          A base

Thus amino acids and proteins are good buffers.

Each protein has a specific pH at which all its molecules are isoelectric, that is, electrically neutral, in solution. That pH is called the *isoelectric point* of the protein. At its isoelectric point a protein is the least soluble. On the acid side (low pH) of the isoelectric point, the protein molecule is positively charged; on the alkaline side (high pH), it is negatively charged.

DENATURATION AND PRECIPITATION OF PROTEINS. Denaturation is a disorganization of the helix structure of the protein molecule due to the breaking up of the cross-links. It occurs when proteins are exposed to denaturing agents and conditions such as the action of acids, bases, alcohol, ultraviolet light, heat, and agitation. Although denaturing agents do not break the peptide linkages, they destroy the biological nature and activity of the protein. Denatured proteins are less soluble but hydrolyze more readily than the original protein.

Denaturation involves precipitation or coagulation of the protein. However, if precipitation is done carefully on certain proteins by the use of certain reagents, denaturation will not take place and the proteins can be redissolved in water without any change in properties. This is desirable in isolating a protein from a solution for study.

Proteins may be precipitated or coagulated in a number of ways:

*By Heat.* Most proteins are coagulated by heat. The cooking of eggs and the coagulation of bacterial protein by sterilization are common examples.

*By Ethyl Alcohol.* Alcohol coagulates nearly all proteins including bacteria. A 70 percent alcohol solution is generally used for antiseptic purposes.

*By Inorganic Acids and Bases.*  The casein of milk is precipitated by the hydrochloric acid of the gastic juice.  Heller's ring test for albumin in the urine is based on the fact that the concentrated nitric acid used precipitates the protein in the form of a white ring.

*By Salts of Heavy Metals.*  When compounds of heavy metals such as silver, lead, copper, and mercury are taken internally, egg white or milk can be given as an antidote to precipitate the heavy metal.  The precipitate formed must be removed by emetic or stomach pump.  However, the best first aid antidote for silver is common salt and that for lead is Epsom salt.

The use of silver nitrate and mercuric chloride as disinfectants is based on their ability to coagulate bacterial protein.

*By Alkaloidal Reagents.*  The alkaloidal reagents are so-called because they precipitate alkaloids.  They can also precipitate proteins.  These reagents include trichloroacetic acid, phosphotungstic acid, tannic acid, and picric acid.  The latter two have been used in the treatment of burns.  They precipitate the surface proteins and thus form a protective coating.

*By Salting Out.*  Proteins may be isolated from solutions by adding neutral salts such as ammonium sulfate, sodium sulfate, magnesium sulfate, or sodium chloride.

HYDROLYSIS.  When treated with acids, alkalies, or protein-splitting enzymes (proteases), proteins are hydrolyzed stepwise as follows:

$$\text{proteins} \xrightarrow{H_2O} \text{proteoses} \xrightarrow{H_2O} \text{peptones} \xrightarrow{H_2O}$$

$$\text{polypeptides} \xrightarrow{H_2O} \text{dipeptides} \xrightarrow{H_2O} \text{amino acids}$$

An example of the hydrolysis of a dipeptide is shown below.

$$CH_3-\underset{\underset{NH_2}{|}}{CH}-\overset{\overset{O}{\|}}{C}-\underset{}{N}\overset{\overset{H}{|}}{}-CH_2-COOH + H_2O \rightarrow$$

Alanylglycine
(dipeptide)

$$CH_3-\underset{\underset{NH_2}{|}}{CH}-COOH + CH_2-COOH$$
$$\underset{NH_2}{|}$$

Alanine                          Glycine
(amino acids)

**Color Reactions of Proteins.**   A number of color tests have been devised to detect the presence of proteins.   The reagents added produce characteristic colors with certain amino acid residues or linkages.

BIURET TEST.   When a few drops of a dilute copper sulfate solution are added to a protein solution containing sodium hydroxide, a violet color is produced.   Biuret ($H_2N$—CO—NH—CO—$NH_2$) also gives a positive test.   This test indicates the presence of peptides containing at least two peptide linkages.

XANTHOPROTEIC TEST.   Concentrated nitric acid produces a yellow color with protein.   The yellow color turns to orange on addition of an alkali.   A positive test is given by amino acids containing the benzene ring.

MILLON TEST.   When a protein solution is heated with Millon's reagent, which is a mixture of mercurous and mercuric nitrates in nitric acid, a red precipitate forms.   This test depends on the presence of tyrosine in the protein.

HOPKINS-COLE TEST.   If a protein solution is mixed with glyoxylic acid (CHO—COOH) and layered over concentrated sulfuric acid, a violet-colored ring will form at the junction of the two liquids.   This test depends on the presence of tryptophan in the protein.

NINHYDRIN TEST.   When a protein containing alpha amino acid is treated with the ninhydrin reagent, a blue color is produced.   This test depends upon the presence of at least one free amino and one free carboxy group in the protein or amino acid.

# 20. NUCLEOPROTEINS AND NUCLEIC ACIDS

Nucleoproteins are found in all living cells, in the cytoplasm as well as in the nucleus. They are the chief constituents of genes, the carriers of hereditary characteristics, and viruses, which cause such infectious diseases as smallpox, mumps, and influenza.

When nucleoproteins are hydrolyzed, they yield nucleic acids and protein (protamine or histone). When nucleic acids are hydrolyzed, they yield phosphoric acid, purine and pyrimidine bases, and a pentose sugar.

Nucleoprotein

   ↓ $H_2O$ + catalyst (acid, alkali, salt, or enzyme)

Nucleic acids + protein

   ↓ $H_2O$ + nuclease

Nucleotides

   ↓ $H_2O$ + nucleotidase

Nucleosides + $H_3PO_4$

   ↓ $H_2O$ + nucleosidase

Purines + pyrimidines + pentose sugar

**Pentose Sugars.** There are two kinds of pentose sugar in nucleic acids and thus two types of nucleic acid, named for the sugar which is present. Thus *ribose* is the sugar in ribonucleic acid (RNA) and *deoxyribose* is the sugar in deoxyribonucleic acid (DNA). The structures of these two sugars are given below.

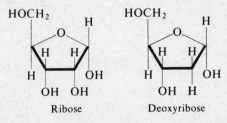

Ribose        Deoxyribose

**Pyrimidine Bases.** The pyrimidine bases found in nucleic acids are *cytosine, uracil,* and *thymine.* Cytosine and uracil are found in RNA. Cytosine and thymine are found in DNA.

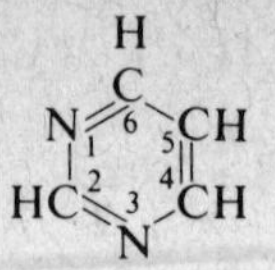

Pyrimidine

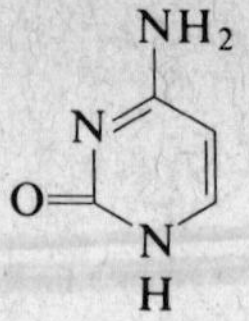

Cytosine
(2-oxy-6-amino pyrimidine)

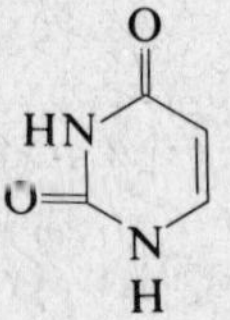

Uracil
(2,6-dioxy pyrimidine)

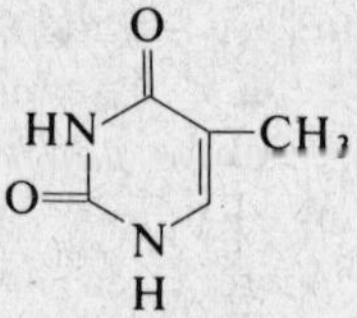

Thymine
(5-methyl uracil)

**Purine Bases.** The two purine bases *adenine* and *guanine* are found in both RNA and DNA.

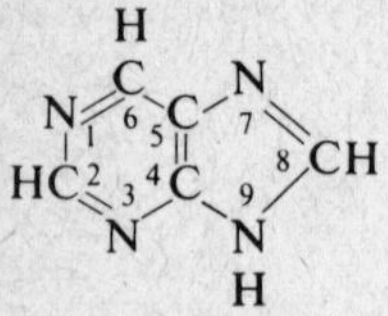

Purine

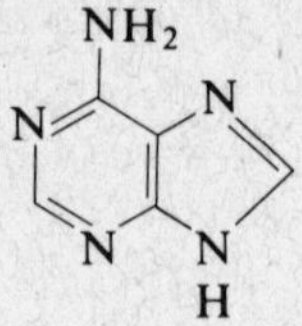

Adenine
(6-amino purine)

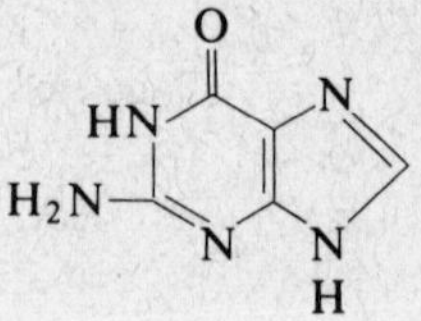

Guanine
(2-amino-6-oxy purine)

**Nucleosides.** When carbon-1 of ribose or deoxyribose is linked to nitrogen of a purine or pyrimidine base, the resulting molecule is called a nucleoside. The five common nucleosides are adenosine

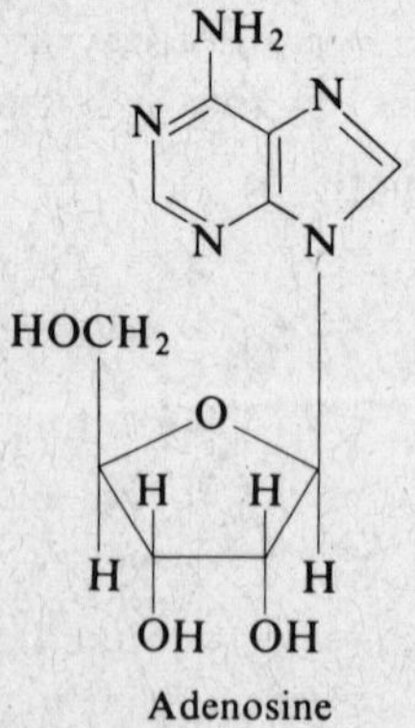

Adenosine

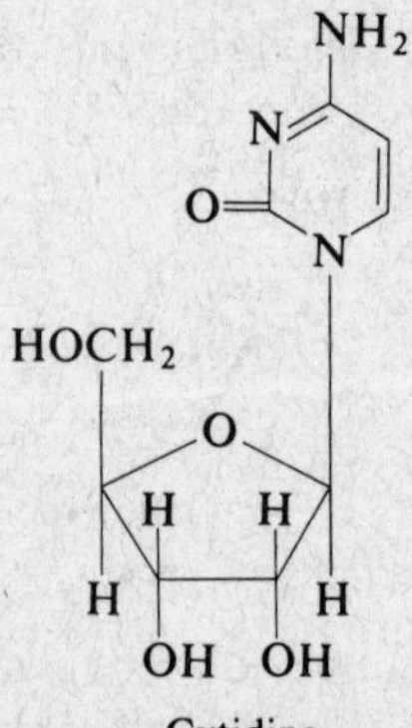

Cytidine

from adenine, guanosine from guanine, cytidine from cytosine, uridine from uracil, and thymidine from thymine. The structural formulas for two of these nucleosides, adenosine and cytidine, are given on p. 124.

**Nucleotides.** When phosphoric acid is linked to a nucleoside, the resulting compound is called a nucleotide. The five common nucleotides are adenylic acid, guanylic acid, cytidylic acid, uridylic acid, and thymidylic acid. Adenylic acid, also known as *adenosine monophosphate* (AMP), is found in muscle tissue. Its two derivatives, *adenosine diphosphate* (ADP) and *adenosine triphosphate* (ATP), are sources of high-energy phosphate bonds and play a vital role in the storage of energy released in the cell during the oxidation of foodstuffs. The structural formula for adenosine triphosphate is shown below. The wavy lines indicate the high-energy phosphate bonds and the vertical broken lines show the relationship between ATP, ADP, and AMP.

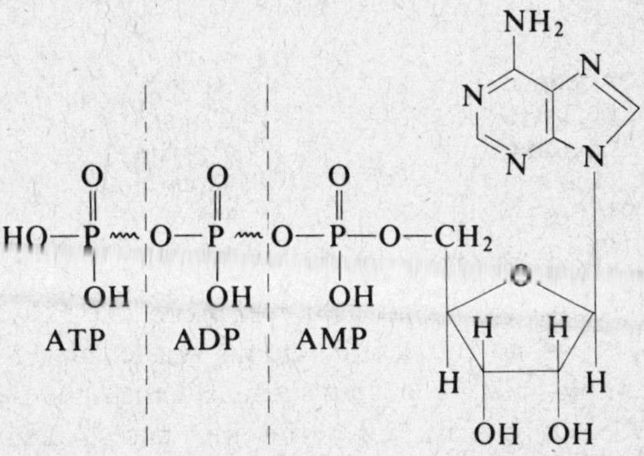

Adenosine triphosphate

Other biologically important nucleotides include the following:

> Nicotinamide adenine dinucleotide (Coenzyme I: NAD; DPN)
> Nicotinamide adenine dinucleotide phosphate (Coenzyme II;
>   NADP; TPN)
> Flavin mononucleotide (FMN)
> Flavin adenine dinucleotide (FAD)

**Nucleic Acids.** Nucleic acids are high molecular weight polynucleotides, being made of many mononucleotides joined together by phosphate groups attached to the pentose groups. A portion of a nucleic acid molecule is shown on p. 126.

$$
\begin{array}{c}
O \qquad\quad \text{Purine} \\
\| \qquad\quad \text{base} \\
O = P - O - \text{pentose} \\
\| \qquad\qquad | \\
OH \qquad\quad O \qquad\quad \text{Pyrimidine} \\
\qquad\qquad\qquad\qquad \text{base} \\
O = P - O - \text{pentose} \\
\| \qquad\qquad | \\
OH \qquad\quad O \qquad\quad \text{Purine} \\
\qquad\qquad\qquad\qquad \text{base} \\
O = P - O - \text{pentose} \\
\| \qquad\qquad | \\
OH \qquad\quad O \qquad\quad \text{Pyrimidine} \\
\qquad\qquad\qquad\qquad \text{base} \\
O = P - O - \text{pentose} \\
\| \qquad\qquad | \\
OH \qquad\quad O \\
\qquad\qquad \text{etc.}
\end{array}
$$

The composition of the two types of nucleic acids is as follows:

| RNA | DNA |
|---|---|
| Adenine | Adenine |
| Guanine | Guanine |
| Cytosine | Cytosine |
| Uracil | Thymine |
| Ribose | Deoxyribose |
| Phosphoric acid | Phosphoric acid |

According to the *Watson-Crick hypothesis*, the DNA molecule is made up of two polynucleotide chains twisted upon each other to form a double helix. The chains are held together by hydrogen bonds between the purine and pyrimidine bases, adenine always being linked with thymine and guanine always with cytosine. The double helix is shown in Fig. 20.1.

**The Genetic Code.** The chromosomes, which are located in the cell nuclei, are strands of DNA and protein. The genes are regions in the DNA strands that contain coded information for the synthesis of a specific protein or an enzyme. As the storage place for genetic information, DNA does not leave the cell nucleus. To transmit information to the sites of protein synthesis in the cytoplasm of the cell, an RNA molecule is formed corresponding to the DNA molecule except that it contains ribose instead of deoxyribose. In the formation of this RNA the DNA strands split apart and serve as templates for the RNA molecule. This RNA, called *messenger RNA*, passes through the nuclear

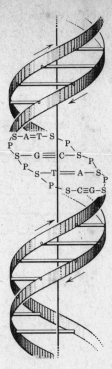

FIG. 20.1. Double helix of DNA. P stands for phosphate diester, S stands for deoxyribose sugar. A=T is the adenine-thymine pairing through two hydrogen bonds and G≡C is the guanine-cytosine pairing through three hydrogen bonds. From *Outlines of Biochemistry* by Conn and Stumpf, 1963, John Wiley & Sons, Inc.

membrane into the cytoplasm where it becomes attached to a ribosome, a nucleoprotein. There it serves in turn as a template, this time for protein synthesis.

The amino acids used to make the protein are brought to the messenger RNA by another RNA called *transfer RNA*. It is thought that there is one transfer RNA for each of the 20 amino acids used in protein synthesis.

The code for the amino acids is carried by the four bases in messenger RNA. (RNA contains the base uracil instead of thymine; uracil can pair with adenine as thymine does.) Current research has indicated that the code "word" for one amino acid is a triplet unit of three of the four bases. In protein synthesis a series of transfer RNA's attach themselves along the messenger RNA chain according to the pairing of the base triplets. The amino acids, which are carried by the transfer RNA's, thus line up in a certain sequence and, as a result of enzyme action, join together by the formation of peptide bonds. A specific protein is the result. The attachment of transfer RNA molecules to a molecule of messenger RNA is shown diagrammatically in Fig. 20.2.

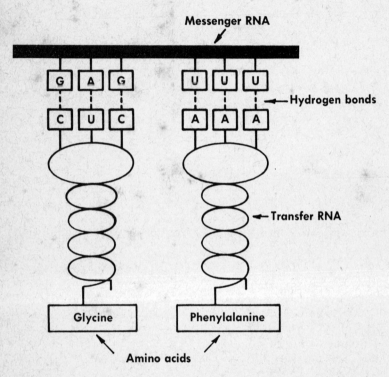

FIG. 20.2. The attachment of transfer RNA molecules to a molecule of messenger RNA. From Arnow: *Introduction to Physiological and Pathological Chemistry*, 7th ed., St. Louis, 1966, The C. V. Mosby Company.

A single error in the "reading" of the code can cause a change in the amino acid sequence which may have serious effects on the functioning of the organism. For example, replacement of one glutamic acid side chain by valine in the hemoglobin molecule results in sickle-cell anemia.

# 21. ENZYMES

None of the chemical reactions taking place in the human body, such as digestion, respiration, and metabolism, would be possible without enzymes, which are the most important tools of the living cell. It has been estimated that there are as many as 1,000 separate enzymes in a single cell, each responsible for a specific chemical reaction. The absence or inaction of a single enzyme can disrupt a key process and result in dysfunction or death of the organism.

**Chemical Nature.** Enzymes are organic catalysts formed by living cells, but their actions are independent of the presence of living cells.

All enzymes have been found to be protein in nature. While some enzymes consist entirely of proteins, others require non-protein components in order to be active. In these enzymes the inactive protein component is known as the *apoenzyme*. The component firmly attached to the enzyme that makes it active is called the *prosthetic group*; it is said to occupy the *active center* or *active site*. If the activating component is easily separated, as by dialysis, it is known as a *coenzyme*. The combined apoenzyme and coenzyme is sometimes referred to as the *holoenzyme*.

Several vitamins of the B complex group have been found to be constituents of certain coenzymes. The most important coenzymes are:

Nicotinamide adenine dinucleotide (Coenzyme I; NAD; DPN)
Nicotinamide adenine dinucleotide phosphate (Coenzyme II; NADP; TNP)
Flavin mononucleotide (FMN)
Flavin adenine dinucleotide (FAD)
Coenzyme A (CoA)

The first four coenzymes act as prosthetic groups of various dehydrogenases and are therefore involved in oxidation-reduction reactions. NAD and NADP accept hydrogen atoms from a substrate and transfer them to FMN and FAD, and FMN and FAD transfer the hydrogen atoms to the cytochromes. The cytochromes in turn transfer the hydrogen atoms to cytochrome oxidase, which activates oxygen so that it may combine with the hydrogen to form water.

Coenzyme A is involved in the metabolism of carbohydrates (citric acid cycle) and the biological syntheses and degradations of fatty acids.

Some enzymes are first produced in an inactive form. Such a precursor of an active enzyme is called a *proenzyme* or *zymogen*. The proenzyme must be activated by some other substance. For example, the inactive proenzyme pepsinogen is converted into active pepsin by the hydrochloric acid of the gastric juice.

**Nomenclature.** Enzymes are usually named by adding the suffix *-ase* to the root of the name of the *substrate*, which is the substance acted upon by the enzyme. For example, *sucrase* hydrolyzes sucrose. Types of enzymes are named according to the nature of the reaction, for example, hydrolases, oxidases, decarboxylases. Many of the early recognized enzymes such as ptyalin, pepsin, and steapsin are still designated by the old suffix *-in*.

**Mode of Action.** The generally accepted theory for enzyme action is that the enzyme (E) first combines with the substrate (S) to form an enzyme-substrate complex (ES). The enzyme-substrate complex undergoes a chemical change and then dissociates, yielding the reaction products and liberating the original enzyme.

$$E + S \rightarrow ES \rightarrow E + Products$$

Enzyme specificity has been explained by the *lock and key theory*, which postulates that each enzyme has an active catalytic center of precise chemical structure or surface shape to which the substrate fits perfectly.

Enzymes catalyze a reaction by lowering the activation energy required to initiate it. Without enzymes chemical reactions in the body would proceed too slowly to maintain life.

**Differences Between Enzymes and Inorganic Catalysts.** Unlike inorganic catalysts, such as platinum, which catalyze many reactions, enzymes are highly specific in their action. Thus lipase will catalyze the hydrolysis of lipids, but not of carbohydrates or proteins. Sucrase will hydrolyze sucrose, but not lactose or maltose.

Enzymes are also more efficient than inorganic catalysts. Sucrase, for example, is 1,000,000 times more powerful than the hydrogen ion in the hydrolysis of sucrose.

Another difference between enzymes and inorganic catalysts is that enzymes are easily destroyed by heat, while inorganic catalysts are not affected by high temperatures.

**Factors that Influence the Rate of Enzyme Action.** Among the factors influencing the rate of enzyme action are the concentration of the substrate and of the enzyme, the temperature, the pH, and the accumulation of end products.

CONCENTRATION OF SUBSTRATE. With an increase in the concentration of the substrate, the reaction rate increases to a maximum or saturation value.

CONCENTRATION OF ENZYME. Within limits, the reaction rate is directly proportional to the concentration of the enzyme.

TEMPERATURE. The optimum temperature of the enzymes present in the body is body temperature, 37°C. (98.6°F.). While each 10°C. rise in temperature usually doubles or triples the reaction rate, rapid inactivation of the enzyme occurs at temperatures much above body temperature. Low temperatures slow down the action of enzymes but do not destroy them.

pH. Each enzyme has its specific optimum pH, at which it exerts its maximum catalytic effect. It loses activity rapidly on either side of this optimum. The optimum pH of several common enzymes is: pepsin 1.5 to 2.2, lactase 5.7, trypsin 7.8.

ACCUMULATION OF END PRODUCTS. Since enzyme reactions are reversible, the accumulation of end products will slow the reaction according to the law of mass action. Some end products affect the pH of the reaction mixture by their acid or alkaline nature, causing the velocity of the reaction to drop.

**Accelerators and Inhibitors.** Some enzymes require the presence of a metallic ion for their activity. Ions which accelerate enzyme action are $Fe^{+2}$, $Fe^{+3}$, $Co^{+2}$, $Zn^{+2}$, $Mn^{+2}$, $Mg^{+2}$, and $Mo^{+2}$.

Certain chemicals have a toxic or inhibitory effect upon enzyme activity. Among these substances are formaldehyde, chloroform, carbon tetrachloride, arsenic compounds, cyanides, and heavy metals like mercury and silver.

Other inhibitors include *antibiotics* such as streptomycin and aureomycin, *antienzymes* such as antitrypsin from soybeans, and *antimetabolites* such as sulfanilamide. The structure of sulfanil-amide closely resembles that of *p*-aminobenzoic acid, which acts as a coenzyme for the synthesis of folic acid in bacteria. When the coenzyme of the bacteria accepts sulfanilamide, the enzyme re-action is inhibited, thus preventing the synthesis of folic acid.

$$NH_2 \qquad\qquad NH_2$$

$$COOH \qquad\qquad SO_2NH_2$$

*p*-Aminobenzoic acid    Sulfanilamide

**Classification.** There have been many different classifications of enzymes, based chiefly on the type of chemical reactions cata-lyzed. The one recommended by the Enzyme Commission of the International Union of Biochemistry contains the following six major classes and a large number of subclasses:

1. Oxidoreductases—catalyze reactions of oxidation and reduction.

2. Transferases—catalyze the transfer of a chemical group from one molecule to another.

3. Hydrolases—catalyze the reaction of hydrolysis.

4. Lyases—catalyze the addition or removal of some chemical group of a substrate, without hydrolysis, oxidation, or reduction.

5. Isomerases—catalyze reactions involving isomerization.

6. Ligases or synthetases—catalyze the joining together of two molecules, coupled with the cleavage of a pyrophosphate bond in ATP or similar compound.

The two major types of reactions catalyzed by enzymes are *hydrolysis* and *oxidation-reduction reactions*. Hydrolytic reactions are concerned mainly with digestion. They are further classified according to the type of food material on which they act. Thus *carbohydrases* catalyze the hydrolysis of starch and sugars, *pro-teases* the hydrolysis of protein and protein derivatives, and *esterases* the hydrolysis of esters such as simple fats and those of phosphoric acid. Oxidation-reduction reactions are concerned with the production of heat and energy in the body. The enzymes involved include the *oxidases*, which activate molecular oxygen; the *dehydrogenases*, which catalyze the removal of hydrogen of a substrate to an easily reducible substance; and the *peroxidases*, which catalyze the decomposition of organic peroxides and hydrogen peroxide.

Table 21.1 lists the principal enzymes involved in hydrolysis and oxidation-reduction reactions.

TABLE 21.1. CLASSIFICATION OF ENZYMES

| Enzyme | Substrate | End Products |
|---|---|---|
| **HYDROLYTIC ENZYMES (HYDROLASES)** | | |
| *Carbohydrases* | | |
| Amylases | Starch | Dextrins and maltose |
| Maltase | Maltose | Glucose |
| Sucrase | Sucrose | Glucose and fructose |
| Lactase | Lactose | Glucose and galactose |
| *Proteases* | | |
| Proteinases | Proteins | Proteoses and peptones |
| Aminopolypeptidase | Polypeptides (free $-NH_2$ end) | Smaller peptides and amino acids |
| Carboxypolypeptidase | Polypeptides (free $-COOH$ end) | Smaller peptides and amino acids |
| Dipeptidase | Dipeptides | Amino acids |
| *Esterases* | | |
| Lipases | Fats | Fatty acids and glycerol |
| Phosphatases | Esters of phosphoric acid | Phosphoric acid and alcohols |
| **OXIDATION-REDUCTION ENZYMES (OXIDO-REDUCTASES)** | | |
| Cytochrome oxidase | Reduced cytochrome C in presence of oxygen | Oxidized cytochrome C and water |
| Succinic dehydrogenase | Succinic acid | Fumaric acid |
| Peroxidase | Phenols and aromatic amines and hydrogen peroxide | Oxidation product of substrate and water |
| Catalase | Hydrogen peroxide | Water and oxygen |
| **OTHER ENZYMES** | | |
| Phosphorylase | Starch or glycogen and phosphate | Glucose-1-phosphate |
| Phosphohexoisomerase | Glucose-6-phosphate | Fructose-6-phosphate |
| Amino acid decarboxylase | Amino acids | Amines and carbon dioxide |
| Transaminase | Glutamic acid and oxalacetic acid, etc. | $\alpha$-Ketoglutaric acid and aspartic acid, etc. |
| Carbonic anhydrase | Carbonic acid | Carbon dioxide and water |
| Zymase | Hexose | Alcohol and carbon dioxide |

# 22. DIGESTION AND ABSORPTION

Our food consists mainly of fats, carbohydrates, proteins, minerals, and vitamins. While minerals and vitamins can pass through the intestinal wall, fats, carbohydrates, and proteins are not able to do so because of their complex nature.

## DIGESTION

Digestion is a process in which complex food material is changed into simple molecules which can be absorbed by the blood from the small intestine and eventually carried to the various tissues where they are needed. It is accomplished by specific enzymes which hydrolyze carbohydrates to monosaccharides, fats to glycerol and fatty acids, and proteins to amino acids.

Digestion begins in the mouth. As the food is chewed and mixed with saliva, the enzyme ptyalin begins to hydrolyze starch. This action continues for fifteen to thirty minutes in the stomach after the swallowing of the food, until the enzyme is inactivated by the hydrochloric acid of the gastric juice.

In the stomach, the mixing of the food with gastric juice is aided by the wavelike contraction of the stomach musculature. The digestion of proteins begins here. The food usually remains in the stomach from two to five hours, depending on the type of food eaten. The thick liquid mixture of partially digested food and gastric juice is called *chyme*.

Chyme passes into the upper portion of the small intestine called the *duodenum*, where its acid is neutralized by the alkaline digestive juices of the pancreas, small intestine, and bile. This mixture also completes the digestion of the partly digested food.

The digestive system is illustrated in Fig. 22.1.

**Salivary Digestion.** Saliva is secreted by three pairs of glands, the parotid, the submaxillary, and the sublingal. The daily output of saliva is approximately 1,500 ml.

Saliva is composed of 99.5 percent of water and 0.5 percent of solids which include inorganic salts, ammonia, urea, uric acid,

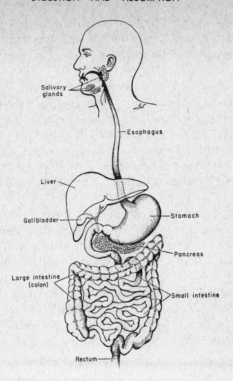

FIG. 22.1. The digestive system. From Brooks: *Basic Facts of General Chemistry*. Philadelphia, W. B. Saunders Co., 1956.

mucin (a glycoprotein which makes the saliva slippery), and the enzyme ptyalin. Saliva has a pH of about 6.8 and a specific gravity of 1.007.

The flow of saliva is stimulated by the following agents: mechanical (mastication or chewing); chemical (taste, especially that of acids); and psychic (sight and smell of food).

The main function of saliva is to begin the digestion of starch. Its principal enzyme, *ptyalin* or salivary amylase, hydrolyzes starch to maltose by the following steps:

Starch + Water → Soluble starch
Soluble starch + Water → Erythrodextrin + Maltose
Erythrodextrin + Water → Achrodextrin + Maltose
Achrodextrin + Water → Maltose

Saliva also has the function of moistening and lubricating the food for swallowing.

**Gastric Digestion.**    The chief purpose of gastric digestion is to break down proteins. This is done by the combined action of the hydrochloric acid and the enzymes that are present in the gastric juice.

GASTRIC JUICE.    The digestive fluid of the stomach is the gastric juice. It contains the enzymes pepsin, rennin, and lipase and also hydrochloric acid, mucin, and some inorganic salts. The proenzyme *pepsinogen* is secreted by the chief cells in the gastric mucosa and the hydrochloric acid by the parietal cells. The hydrochloric acid is believed to be formed in the cells from carbonic acid and sodium chloride. The pH of gastric juice is 1.6 to 1.8.

The daily output of gastric juice in a normal person is 2,000 to 3,000 ml. Sight, smell, taste, or the thought of appetizing foods stimulate its flow. The gastric secretory glands are also stimulated by the hormone-like substance *gastrin*, which is formed in the lower or pyloric portion of the stomach. The production of gastrin is initiated by the presence of food in the stomach.

*Hydrochloric acid* is present in the gastric juice in the range of 0.2 to 0.5 percent. It acts as a germicide for pathogenic bacteria swallowed with food as well as providing the correct pH for gastric digestion. It also activates the enzyme pepsinogen.

*Pepsin* is secreted in the inactive zymogen form, *pepsinogen*, which is converted into active pepsin by hydrochloric acid. Pepsin begins the digestion of proteins by hydrolyzing the peptide bonds to produce proteoses and peptones.

*Rennin*, like pepsin, is a protease. It changes casein, the main protein in milk, to paracasein, which combines with calcium to form the insoluble calcium paracaseinate curd. The milk curds are then acted upon by pepsin. If milk were not first reduced to curd, it would pass out of the stomach too quickly to be digested. Rennin is abundant and active in infants and young animals. There is some evidence that little or no rennin exists in the gastric juice of adults. Commercial rennin is used in cheese-making and in the preparation of junket.

*Lipase*, the fat-splitting enzyme, is present in gastric juice, but very little digestion of fat takes place in the stomach. Those fats that are partially digested by gastric lipase are milk and egg yolk, which are highly emulsified. It is not known whether the lipase in the stomach is secreted with the gastric juice or enters from the small intestine.

GASTRIC ANALYSIS. The presence or absence of many substances in the gastric juice, as well as abnormally high or abnormally low values of the regular constituents, can be determined by analysis of a withdrawn sample. The information obtained may reveal disease conditions of the individual. *Achlorhydria* is the absence of free hydrochloric acid, a condition commonly associated with pernicious anemia or stomach cancer. In *hypochlorhydria* (hypoacidity), there is an abnormally low concentration of acid, an indication of stomach cancer, chronic constipation, or inflammation of the stomach. In *hyperchlorhydria* (hyperacidity), conversely, the stomach contains excess acid, usually due to peptic or duodenal ulcers, or inflammation of the gallbladder (cholecystitis).

**Intestinal Digestion.** The small intestine has three digestive juices: pancreatic juice, intestinal juice, and bile.

PANCREATIC JUICE. Pancreatic juice is secreted by the pancreas. It reaches the intestine through the pancreatic duct which joins the common bile duct before opening into the duodenum.

The daily output of pancreatic juice is approximately 650 ml. It is stimulated by the hormone *secretin* which is liberated from the intestinal mucosa by the acid present in the chyme.

Pancreatic juice is composed of 98.7 percent of water and 1.3 percent solids consisting of digestive enzymes, albumin, globulin, proteoses, peptones, sodium chloride, and sodium bicarbonate. It has a pH of about 8. It contains enzymes capable of hydrolyzing all three classes of foods.

*Trypsin* is secreted as the precursor *trypsinogen*, which is activated to trypsin by *enterokinase* present in the intestinal juice. It hydrolyzes proteins, proteoses, and peptones to polypeptides and amino acids.

*Chymotrypsin* is secreted as the precursor *chymotrypsinogen*, which is activated to chymotrypsin by trypsin. Like trypsin, it hydrolyzes proteins, proteoses, and peptones to polypeptides and amino acids. It also acts on casein in a manner similar to that of rennin.

*Carboxypolypeptidase* hydrolyzes polypeptides to simpler peptides and amino acids. Starting at the end of the chain, where there is a free carboxyl group, it splits off the amino acid to which this group is attached, and then splits off the next one, and the next, until the chain has been completely hydrolyzed.

*Pancreatic amylase* (amylopsin) hydrolyzes starch to maltose in the same manner as does ptyalin, but with greater digestive power.

*Pancreatic lipase* (steapsin) hydrolyzes fats into glycerol and fatty acids. It is responsible for most of the fat digestion in the body.

INTESTINAL JUICE. The intestinal juice (succus entericus) is secreted by the intestinal mucosa. It completes the digestion of the proteins and the disaccharides and thus contains peptidases and disaccharases. *Dipeptidase* hydrolyzes dipeptides to amino acids. *Sucrase*, *lactase*, and *maltase* hydrolyze the respective disaccharides to glucose and other monosaccharides. The intestinal juice also contains *lecithinase* for the digestion of lecithin and *nucleases* for the digestion of nucleic acids.

BILE. Bile is a brownish-yellow or greenish-yellow alkaline fluid which is continuously formed in the liver and stored and concentrated in the gallbladder. When partially digested fatty food enters the duodenum, the hormone *cholecystokinin* is liberated into the bloodstream. This hormone causes the gallbladder to contract and empty the bile into the intestine. The normal person produces between 500 and 800 ml. of bile per day.

Bile is not a digestive juice because it contains no enzymes; however, it serves an important digestive function by emulsifying fats and preparing them for the action of steapsin. It also aids in the excretion of certain waste products.

Bile consists of water, salts, pigments, cholesterol and other lipids, mucin, and inorganic salts.

*Bile Salts.* The two principal bile salts are *sodium glycocholate* and *sodium taurocholate*. Sodium glycocholate is the sodium salt of glycocholic acid (cholic acid + glycine), and sodium taurocholate is the sodium salt of taurocholic acid (cholic acid + taurine). See structural formulas on p. 139.

The chief function of the bile salts in digestion is the emulsification of fats by lowering the surface tension of water. They also aid in the absorption of fatty acids through the intestinal wall.

*Bile Pigments.* The principal bile pigment is *bilirubin*, a red pigment which, along with its oxidized form *biliverdin* (green), is formed in the liver from disintegrated hemoglobin of the red blood cells. Both of these pigments are excreted in the bile. Bilirubin is reduced by intestinal bacteria to *stercobilin*, the normal brown pigment of the feces. Bilirubin is also reduced to

$$CH_2-CH_2-SO_3H$$

$$H_2N-CH_2-COOH \qquad NH_2$$

Glycine                            Taurine

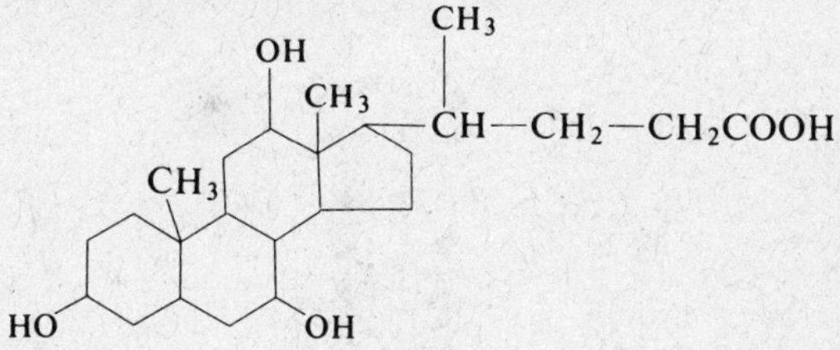

Cholic acid

*urobilin* and *urobilinogen*, which mixture is called *urochrome*, the chief pigment of the urine.

Hemoglobin
| hydrolysis
Biliverdin (bile) + iron + globin
| reduction
Bilirubin (bile)
| reduction
Urobilinogen (intestine) $\xrightarrow{\text{oxidation}}$ urobilin (urine)
| reduction
Stercobilin (feces)

Jaundice is a condition in which there is a yellowing of the skin and the sclera (white portion) of the eyes due to an abnormal amount of bilirubin in the blood. *Obstructive jaundice* is caused by an obstruction in the common bile duct, while *hemolytic jaundice* is the result of excessive destruction of red cells.

*Cholesterol.* Cholesterol is excreted from the body through the bile. Occasionally, in the presence of foreign substances (injured cells, bacteria, or pigments), cholesterol precipitates in the form of gallstones.

## ABSORPTION

The passage of the end products of digestion from the small intestine into the bloodstream is called *absorption*.

The small intestine, which is approximately 25 feet long, has many folds in its mucous lining. Its absorbing surface is further increased by the presence of from 4,000,000 to 5,000,000 finger-like projections called *villi*. These villi present a surface area of about 109 square feet.  Each villus contains a central lymph capillary called a *lacteal*, which is surrounded by a network of blood capillaries.  A diagram of a villus is given in Fig. 22.2.

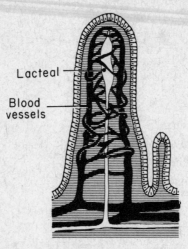

FIG. 22.2.  Diagram of a villus.  From Routh: *20th Century Chemistry*, 3rd ed. Philadelphia, W. B. Saunders Co., 1963.

Digested food normally remains in the small intestine from five to eight hours.

**Absorption of Carbohydrates.**  Carbohydrates are absorbed as monosaccharides, whose different rates of absorption are: galactose > glucose > fructose > mannose.  They are absorbed into the bloodstream through the capillary blood vessels of the villi, then carried to the liver by way of portal circulation.  There they are converted into glycogen for storage or passed on to the blood as glucose which is oxidized in the cells to give heat and energy.

**Absorption of Fats.**  Fats, in the form of mono- and diglycerides, glycerol, and fatty acids, are absorbed through the lacteals of the villi with the aid of bile salts, which are thought to form complexes with them.  The hydrolyzed or partially hydrolyzed glycerides are resynthesized into triglycerides which are carried via the lymph into the thoracic duct and finally into the blood.  The fat carried by the blood is either immediately metabolized or stored in the adipose tissue of the body.

**Absorption of Proteins.** Proteins are absorbed in the form of amino acids. The amino acids, like the monosaccharides, are absorbed directly into the bloodstream through the capillary blood vessels of the villi. They are carried to the tissues to be used for synthesis of new tissue proteins and other nitrogen-containing compounds, or oxidized to produce energy.

## BACTERIAL ACTION IN THE COLON

*Fermentative* bacteria in the colon act upon any undigested carbohydrates to produce gases (carbon dioxide, methane, and hydrogen) and acids (acetic, lactic, and butyric). *Putrefactive* bacteria act upon proteins to produce phenols, indole, skatole, histamine, and other toxic substances. Indole and skatole are responsible for the objectionable odor of the feces. The products of putrefaction, if absorbed, are detoxified by the liver and then excreted. For example, indole is converted to indican and excreted in the urine.

After the absorption of most of the water, the material left in the colon is converted to *feces*. The feces are composed mainly of the skin and seeds of fruits, vegetable fiber, waste material from cellular decomposition, bile pigments, cholesterol, inorganic salts, and living and dead bacteria, which may amount to one-fourth to one-half of the dry weight of fecal matter.

# 23. CARBOHYDRATE METABOLISM

*Metabolism* refers to the chemical changes that the absorbed products of digestion undergo in the tissues of the living body. The two main types of metabolism are *anabolism* and *catabolism*. *Anabolism* is a building-up process whereby simple products of digestion are made into complex molecules to form new tissue, repair old tissue, store food supplies, and synthesize enzymes, pigments, hormones, etc. *Catabolism*, on the other hand, is a breaking-down process in which absorbed products of digestion and worn-out tissues are reduced to simple waste products such as carbon dioxide, water, and urea with the simultaneous release of energy.

**Basal Metabolic Rate.** Basal metabolic rate (BMR) is defined as the amount of heat produced by the body to maintain life processes when the body is in a state of physical, emotional, and digestive rest. It is expressed as kilocalories per hour per square meter of body surface, which is determined from the height and weight of the individual. The basal metabolic rate for a normal adult male is 40, and that for a normal adult female, 37. A variation of $\pm 15$ percent from the standard average value is considered normal.

The determination of basal metabolism is made in the morning after a light evening meal and a full night's rest, by measuring the volume of oxygen consumed in the given time, usually 6 to 8 minutes. (One liter of oxygen consumed corresponds to 4.825 kilocalories of heat metabolized.)

Basal metabolic rate is a valuable diagnostic tool. For example, in hyperthyroidism the basal metabolic rate may be as much as $+80$ percent above normal, and in hypothyroidism it may be as much as $-40$ percent below normal.

**Blood Sugar Level.** The end products of carbohydrate digestion are the monosaccharides glucose, fructose, and galactose. These simple sugars are absorbed into the blood and are carried through the portal vein to the liver where they are converted into glycogen, the reserve carbohydrate in the body.

Glucose is always present in the blood. When the concentration drops slightly below the *normal fasting level* (70 to 90 mg. per 100 ml. of blood), as during normal muscular activity, glycogen stored in the liver is hydrolyzed to glucose, which passes into the blood to maintain the normal sugar level. After strenuous exercise, which uses up glucose faster than it can be converted

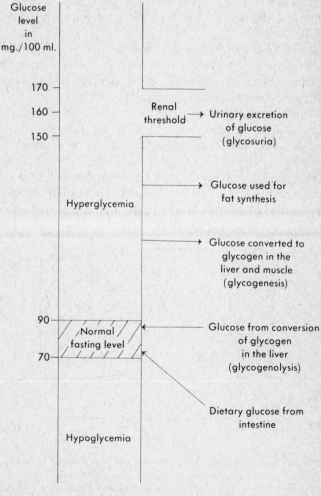

FIG. 23.1. Regulation of blood sugar level.

from glycogen by the liver, blood sugar concentration may temporarily fall below the normal fasting level, causing *hypoglycemia*.

After a heavy meal of carbohydrate food, the monosaccharides are absorbed into the blood faster than they can be converted into glycogen by the liver. The blood sugar level rises, causing a temporary *hyperglycemia*. If the blood sugar level continues to rise, the glucose may be transformed into fat and stored in the adipose tissue. When the blood sugar level reaches the *renal threshold* (150 to 170 mg. per 100 ml.), glucose is excreted by the kidneys and appears in the urine, a condition known as *glycosuria*. See Fig. 23.1.

**Glucose Tolerance Test.** This test is used to confirm a suspected diagnosis of diabetes. It is based on the fact that after the ingestion of one gram of glucose per kilogram of body weight by a fasting person, the following differences are observed:

(1) In a normal person, the blood sugar level rises to 160 to 170 mg. during the first hour, while in the diabetic, the blood sugar level rises to 200 to 300 mg. in the same amount of time depending on the severity of the disease.

(2) In the normal person, the high blood sugar level usually returns to normal before the end of the second hour, while in the diabetic, the blood sugar level remains above the renal threshold for 6 to 12 hours, as shown in Fig. 23.2.

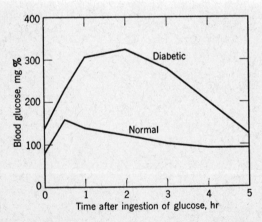

FIG. 23.2. Sugar tolerance curves of normal and diabetic human patients. The data refer to the arterial blood sugar levels after the ingestion of 100 gm. of glucose. From *General Biochemistry* by Fruton and Simmonds, 2nd ed., 1958, John Wiley & Sons, Inc.

**Glycogenesis.** Glycogenesis is the conversion of glucose to glycogen in the liver and in the muscle. The energy required for glycogenesis is derived from the breaking of one of the high-energy phosphate bonds (represented by a wavy line $\sim$ ) in adenosine triphosphate (ATP), resulting in the formation of adenosine diphosphate (ADP) and inorganic phosphate ion, with the release of energy.

$$\text{Adenine-ribose}-\underset{\underset{OH}{|}}{\overset{\overset{O}{\|}}{P}}-O\sim\underset{\underset{OH}{|}}{\overset{\overset{O}{\|}}{P}}-O\sim\underset{\underset{OH}{|}}{\overset{\overset{O}{\|}}{P}}-OH \rightarrow$$

Adenosine triphosphate (ATP)

$$\text{Adenine-ribose}-\underset{\underset{OH}{|}}{\overset{\overset{O}{\|}}{P}}-O\sim\underset{\underset{OH}{|}}{\overset{\overset{O}{\|}}{P}}-OH + PO_4^{-3} + \text{energy}$$

Adenosine diphosphate (ADP)

Glucose is converted into glycogen in the following steps:

I Glucose + ATP $\xrightarrow{\text{hexokinase}}$ Glucose-6-phosphate + ADP

II Glucose-6-phosphate $\underset{\xrightarrow{\hspace{2cm}}}{\text{phosphoglucomutase}}$ Glucose-1-phosphate

III Glucose-1-phosphate $\xrightarrow{\text{glycogen phosphorylase}}$ Glycogen + $PO_4^{-3}$

In step I, the terminal phosphate group lost by ATP in its conversion to ADP is transferred to glucose, forming glucose-6-phosphate. This process is called *phosphorylation*. The enzyme involved in the phosphorylation of glucose is *hexokinase*. The enzyme *phosphoglucomutase* isomerizes glucose-6-phosphate to glucose-1-phosphate, and *glycogen phosphorylase* converts glucose-1-phosphate to glycogen.

Fructose and galactose, the other two monosaccharides entering the liver with glucose, are phosphorylated by ATP and a specific enzyme, then transformed into glucose-6-phosphate. In this way they, too, can be converted to glycogen.

**Glycogenolysis.** Glycogenolysis is the conversion of glycogen to glucose, the reverse of glycogenesis. It takes place only in the liver, not in the muscle. Since step I in the glycogenesis above is not reversible, a different enzyme is required for the conversion of glucose-6-phosphate to glucose in step III below.

I  Glycogen + $PO_4^{-3}$ $\xrightleftharpoons{\text{phosphorylase}}$ Glucose-1-phosphate

II  Glucose-1-phosphate $\xrightleftharpoons{\text{phosphoglucomutase}}$ Glucose-6-phosphate

III  Glucose-6-phosphate + ADP $\xrightarrow{\text{glucose-6-phosphatase}}$ Glucose + ATP

**Oxidation of Glucose.**    The complete combustion of one mole of glucose (180 grams) liberates 686 kilocalories of heat.

$$C_6H_{12}O_6 + 6O_2 \rightarrow 6CO_2 + 6H_2O + 686 \text{ Kcal./mole}$$

However, as a source of energy in the body, glucose is not oxidized directly, in one step, to the end products carbon dioxide and water. Rather, energy is released in small bursts from a series of individual reactions. For each reaction there is a different enzyme, which in many cases also requires specific coenzymes. The energy liberated is stored in the high-energy bonds of a number of molecules of ATP, which is synthesized from ADP molecules and the phosphate ion.

The same amount of energy is released when ATP is hydrolyzed to ADP. Thus ATP serves as a readily available source of energy. It may supply the heat energy for the maintenance of body temperature, mechanical energy for muscular contraction, electric energy for conduction of nerve impulses, or chemical energy for the synthesis of glycogen, hormones, and tissue proteins.

Of these different energy transformations the one that has been most intensively studied is that of muscular contraction. The discussion that follows deals with this phase of energy transformation.

The oxidation of glucose in the muscle may be divided into two phases: the anaerobic or contractile phase, which proceeds in the absence of oxygen, and the aerobic or recovery phase, which requires oxygen.

ANAEROBIC PHASE.    In the anaerobic phase, glucose and glycogen are converted into lactic acid, a process called *glycolysis*. The energy liberated is used for the resynthesis of ATP from ADP and the phosphate ion. The reactions involved are shown in Fig. 23.3.

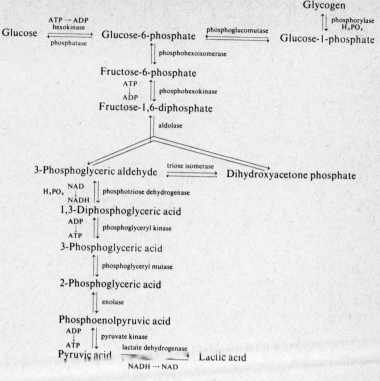

FIG. 23.3. Glycolysis.

The overall reaction in glycolysis may be summarized by the following equation:

$$C_6H_{12}O_6 + 2ADP + 2H_3PO_4 \rightarrow$$
Glucose

$$2CH_3-\underset{\underset{H}{|}}{\overset{\overset{OH}{|}}{C}}-COOH + 2ATP + 2H_2O$$

Lactic acid

The energy for the conversion of ADP to ATP is supplied by glucose indirectly. The immediate energy for this conversion is supplied by the hydrolysis of creatine phosphate, or phosphocreatine. The energy for the reconversion of creatine to creatine phosphate is supplied by the hydrolysis of glycogen to lactic acid.

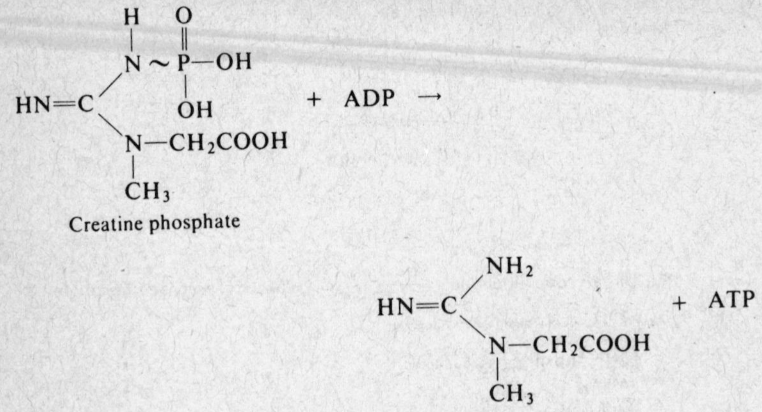

$$HN=C \begin{cases} N \sim P-OH \\ \quad\quad OH \\ N-CH_2COOH \\ CH_3 \end{cases} \quad + \text{ ADP} \rightarrow$$

Creatine phosphate

$$HN=C \begin{cases} NH_2 \\ N-CH_2COOH \\ CH_3 \end{cases} \quad + \text{ ATP}$$

Creatine

AEROBIC PHASE. Four-fifths of the lactic acid formed by glycolysis in the contracting muscle is carried by the blood to the liver where it is reconverted to glycogen. The remaining one-fifth lactic acid is reconverted to pyruvic acid, which is oxidized to carbon dioxide and water, with the release of energy, through what is called the citric acid cycle, also known as the Krebs cycle or the tricarboxylic acid cycle. The citric acid cycle is shown in Fig. 23.4.

The overall reaction in the citric acid cycle may be summarized by the following equation:

$$2CH_3COCOOH + 5O_2 \rightarrow 6CO_2 + 4H_2O + \text{energy}$$
Pyruvic acid

The two moles of pyruvic acid (formed by glycolysis from one mole of glucose) yield 30 moles of ATP. Eight moles of ATP are produced from glycolysis itself; therefore 38 moles of ATP are obtained per mole of glucose oxidized to carbon dioxide and water. This is an overall energy yield of 42 percent.

**Hormones That Influence Carbohydrate Metabolism.** The principal hormones that affect carbohydrate metabolism are insulin, epinephrine, and glucagon.

INSULIN. Insulin is produced by the pancreas. The lack or deficiency of this hormone causes *diabetes mellitus*. Insulin facilitates the transfer of glucose into the cell. It removes glucose

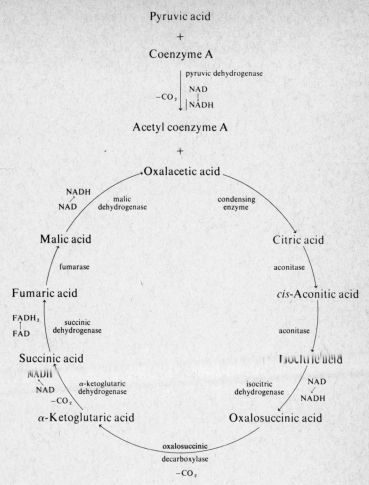

Fig. 23.4.   Citric acid cycle.

from the bloodstream by hastening the conversion of glucose to glycogen in the liver and muscle, by speeding up the oxidation of glucose in the tissues, by inhibiting the breakdown of liver glycogen, and by promoting the formation of fat from glucose.

EPINEPHRINE (ADRENALIN).   Epinephrine is produced by the adrenal glands and discharged into the bloodstream when the individual is under stress.   It accelerates the conversion of liver glycogen to glucose, thereby increasing the blood sugar level and providing quick energy to help the body meet an emergency. Its action is antagonistic to that of insulin.

GLUCAGON. Glucagon is another hormone produced by the pancreas. Like epinephrine, it causes a breakdown of liver glycogen and a rise in blood sugar. Since its action is antagonistic to that of insulin, it has been used to treat insulin intoxication.

# 24. LIPID METABOLISM

Fats have a higher caloric value than either carbohydrates or proteins. They may be oxidized in the tissues to yield energy or stored as reserve fuel under the skin and around many of the body organs. The fat so deposited is called *depot fat* or *adipose tissue*. Besides serving as a reserve fuel, depot fat cushions and supports the vital organs and protects the interior of the body from changes in outside temperature. Excessive storage of depot fat results in obesity.

Fats may also be used in the synthesis of other lipids such as phospholipids and cholesterol esters of fatty acids.

**Oxidation of Fats.** The fats are probably converted to phospholipids before they are oxidized. The phospholipids are then hydrolyzed to glycerol and fatty acids.

GLYCEROL METABOLISM. The glycerol is dehydrogenated and phosphorylated to 3-phosphoglyceraldehyde, which enters the glycolytic pathway to be converted into glycogen or into pyruvate. As pyruvate, it would enter the citric acid cycle and be oxidized to carbon dioxide and water.

$\beta$-OXIDATION OF FATTY ACIDS. The fatty acids are oxidized according to Knoop's beta-oxidation theory. Knoop discovered that a fatty acid loses two carbon atoms at a time during oxidation, and that oxidation takes place at the beta carbon atom.

The fatty acid is first activated by combining with coenzyme A(CoA · SH). See step 1 in sequence of reactions in Fig. 24.1. Then by two dehydrogenation reactions separated by a hydration reaction, the $\beta$-carbon atom (the second carbon atom from the carboxyl group) is oxidized to a keto-acid (step 4). Hydrolysis causes a split between $\alpha$- and $\beta$-carbon atoms, forming the two-carbon fragment acetyl coenzyme A and a new fatty acid that is two carbon atoms shorter than the original acid (step 5). The acetyl coenzyme A is eventually oxidized in the citric acid cycle to carbon dioxide and water. Each step in the sequence of reactions is catalyzed by a different enzyme. Coenzyme A remains

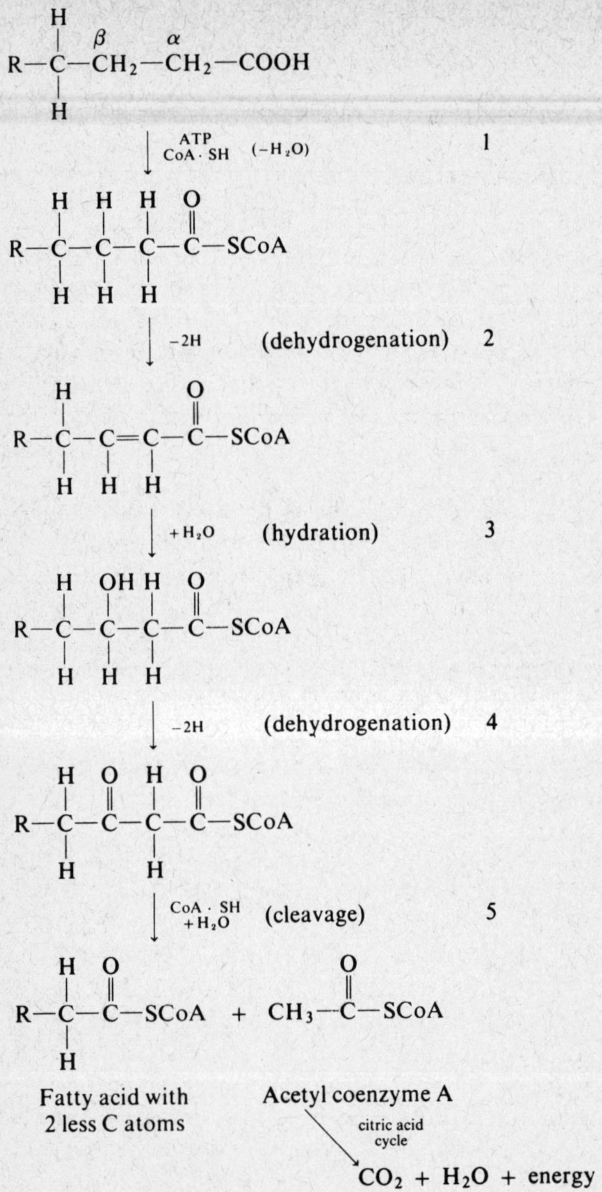

FIG. 24.1   $\beta$-Oxidation of fatty acids.

attached to the fatty acid throughout. (SH, the mercapto group, is the active site on coenzyme A. In reaction with an acid, the H is split off and, with the OH of the acid, forms water. Coenzyme A then joins the acid through the sulfur atom.)

The $\beta$-oxidation process is repeated; each time, the fatty acid is shortened by two carbon atoms until a two-carbon fragment is reached, as shown below without coenzyme A.

$$C_{11}H_{23}COOH \rightarrow C_9H_{19}COOH \rightarrow C_7H_{15}COOH \rightarrow$$
$$C_5H_{11}COOO \rightarrow C_3H_7COOH \rightarrow CH_3COOH \rightarrow CO_2 + H_2O$$

**Ketone Bodies.** The ketone, or acetone, bodies are acetoacetic acid, $\beta$-hydroxybutyric acid, and acetone. They are formed in the liver in the catabolism of fatty acids and are utilized principally by the muscles. Large amounts of ketone bodies are produced when there is restricted carbohydrate metabolism and increased fat metabolism, as occurs in high fatty diet, diabetes mellitus, starvation, or severe liver damage. This condition is called *ketosis*.

Because acetoacetic acid and $\beta$-hydroxybutyric acid are excreted in the urine in the form of sodium or other metallic salts, the alkali reserve of the body becomes depleted when there is an excess of ketone bodies. The result is *acidosis*.

The formation of ketone bodies is shown below.

$$-CO_2 \quad CH_3-CO-CH_3$$
Acetone

$$CH_3-CO-CH_2-COOH$$
Acetoacetic acid

$$+2H$$

$$\overset{H}{\underset{OH}{CH_3-\overset{|}{\underset{|}{C}}-CH_2-COOH}}$$
$\beta$-Hydroxybutyric acid

**Biosynthesis of Fat.** In $\beta$-oxidation, fatty acids are degraded into two-carbon units; they are synthesized from two-carbon units by a modified reverse process. Thus, acetic acid units, through the action of coenzyme A, can be joined to form long-chain fatty acids.

In triglyceride synthesis, glycerol, by reacting with ATP, is converted to $\alpha$-glycerophosphate, which reacts with two molecules of fatty acid coenzyme A to form $\alpha$-phosphatidic acid. The splitting off of the phosphate group from α-phosphatidic acid results in the formation of α,β-diglyceride, which reacts with another molecule of fatty acid coenzyme A to form the triglyceride (fat) and coenzyme A.

Three important fatty acids that the body cannot synthesize are linoleic, linolenic, and arachidonic acids. These are known as *essential fatty acids* and must be supplied by the diet.

**Role of the Liver in Lipid Metabolism.** The liver is the chief site for both degradation and synthesis of fat. Its normal fat content is 4 to 5 percent. In the condition known as *fatty liver*, its fat content may be raised to as much as 50 percent. Fatty liver can be caused by a diet high in fat or cholesterol and low in protein; by starvation; by disease, such as diabetes mellitus; or by poisons, such as phosphorus, chloroform, and carbon tetrachloride. Fatty infiltration of the liver is believed by some to be curable or preventable by the use of *lipotropic factors* or *agents*, such as choline and methionine, which help to form phospholipids, the most important fat carriers.

**Phospholipid Metabolism.** Dietary phospholipids are hydrolyzed in the small intestine by phosphatidases produced in the pancreas. The products are glycerol, fatty acids, phosphate, and nitrogenous bases. Phospholipids are synthesized in the liver.

The body utilizes phospholipids in several ways. Lecithin is involved in fat transport and is a source of phosphoric acid for cell building. Cephalin plays an important role in blood clotting, and sphingomyelin is a main component of the myelin sheath of the nerves.

**Cholesterol Metabolism.** Cholesterol is the chief animal sterol and is found in every cell in the body. It is believed to be the precursor of a variety of compounds, including sex hormones, bile acids or salts, adrenal cortical hormones, gallstones, and vitamin D.

Cholesterol can be synthesized in the body from two-carbon compounds such as acetyl coenzyme A. The liver is the main site of cholesterol synthesis. Cholesterol is eliminated in the bile. It is deposited in the arterial walls in *atherosclerosis*.

# 25. PROTEIN METABOLISM

Proteins are hydrolyzed to amino acids by the action of the proteolytic enzymes of the alimentary tract. The amino acids pass readily through the walls of the intestine into the blood. They are carried by the portal vein to the liver from which they enter the general circulation for distribution to the various body tissues.

**The Amino Acid Pool.** There are no storage depots in the body for proteins or amino acids. The dietary amino acids from the process of absorption and those formed by the body through synthesis or through the breakdown of body proteins make up what is called the *amino acid pool* or the *nitrogen pool*. There is a constant interchange between the amino acids of tissue proteins and those in the amino acid pool.

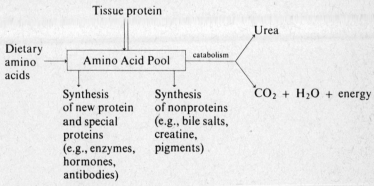

**Nitrogen Equilibrium.** A healthy adult on a proper diet who maintains a constant weight excretes as much protein-nitrogen as he takes in and is said to be in *nitrogen equilibrium* or *nitrogen balance*. A child retaining nitrogen to provide for growth and a person undergoing new tissue formation after a debilitating disease are said to have a *positive nitrogen balance*. Persons suffering from wasting diseases, prolonged fevers, or starvation, where more nitrogen is excreted than is present in the food intake, are said to have a *negative nitrogen balance*.

**Metabolic Reactions of Amino Acids.** The amino acids in the pool can undergo three important metabolic reactions: oxidative deamination, transamination, and urea formation.

OXIDATIVE DEAMINATION. Certain amino acids not needed by the body undergo oxidative deamination to form ammonia and a keto acid. This reaction, which involves the splitting off of the amino group of the amino acid, is catalyzed by an amino acid oxidase.

$$CH_3-\underset{\underset{NH_2}{|}}{CH}-COOH + [O] \xrightarrow[\text{oxidase}]{\text{amino acid}} CH_3-\underset{\underset{O}{\|}}{C}-COOH + NH_3$$

Alanine            Pyruvic acid     Ammonia
                          (keto acid)

TRANSAMINATION. Transamination is a chemical reaction in which an animo group from an amino acid is transferred to a keto acid. The new amino acid produced is related to the original keto acid. Thus new amino acids can be synthesized from the reaction of excess amino acids in the pool with keto acids formed from the deamination of other amino acids.

$$
\begin{array}{cccc}
\text{COOH} & \text{COOH} & \text{COOH} & \text{COOH} \\
| & | & | & | \\
CH_2 & C=O & CH_2 & CH_2 \\
| & | & | & | \\
CH_2 \; + & CH_2 & \underset{\text{transaminase}}{\rightleftarrows} & CH_2 \; + & CHNH_2 \\
| & | & | & | \\
CHNH_2 & COOH & C=O & COOH \\
| & & | & \\
COOH & & COOH & \\
\end{array}
$$

Glutamic    Oxalacetic         α-Ketoglutaric    Aspartic
   acid        acid             acid         acid
(amino acid)   (keto acid)       (keto acid)    (amino acid)

**Formation of Urea.** Most of the ammonia resulting from the deamination and oxidation of the amino acids is converted in the liver into urea, the main end product of protein metabolism. The formation of urea involves a series of reactions known as the *ornithine cycle*. First the ammonia and carbon dioxide combine with the amino acid ornithine to form another amino acid, citrulline. Citrulline combines with another molecule of ammonia to form the amino acid arginine, which is hydrolyzed with the aid of the enzyme arginase to urea and ornithine. Thus the ornithine is recovered for reuse in the cycle. The ornithine cycle is illustrated on p. 157.

$$
\begin{array}{cccc}
& NH_2 & NH_2 & NH_2 \\
& | & | & | \\
& C{=}O & C{=}NH & C{=}O \\
& | & \text{-----}+\text{-----} & | \\
NH_2 & NH & NH & NH_2 + NH_2 \\
| & | & | & \quad Urea \\
CH_2 & CH_2 & CH_2 & CH_2 \\
| & | & | & | \\
CH_2 & CH_2 & CH_2 & CH_2 \\
| & \xrightarrow[CO_2]{NH_3} | & \xrightarrow{NH_3} | & \xrightarrow[\text{arginase}]{H_2O} | \\
CH_2 & CH_2 & CH_2 & CH_2 \\
| & | & | & | \\
CHNH_2 & CHNH_2 & CHNH_2 & CHNH_2 \\
| & | & | & | \\
COOH & COOH & COOH & COOH \\
\text{Ornithine} & \text{Citrulline} & \text{Arginine} & \text{Ornithine}
\end{array}
$$

**Essential Amino Acids.** Essential amino acids are so-called because they cannot be synthesized by the body and must be supplied in the diet. Without them the body is unable to form tissue proteins. The essential amino acids for humans are leucine, isoleucine, phenylalanine, methionine, lysine, valine, threonine, and tryptophan. Arginine and histidine are essential in the rat.

Proteins that contain adequate amounts of all the essential amino acids are called *complete proteins* or *adequate proteins*. The proteins in meat, milk, eggs, and soybeans are complete proteins. Proteins deficient in one or more essential amino acids are called *incomplete proteins*. Examples of incomplete proteins are gelatin, which lacks tryptophan, and zein of corn, which lacks both tryptophan and lysine.

**Creatine and Creatinine.** *Creatine* is synthesized from the amino acids glycine, arginine, and methionine. It is found in muscle as creatine phosphate (phosphocreatine), which furnishes energy for muscular contraction by regenerating ATP from ADP (see equation on p. 158).

*Creatinine* is a waste product derived from creatine present in muscle tissue. The amount of creatinine excreted in the urine by an individual is nearly constant, amounting to 1 to 1.5 grams per day. The equation on p. 158 shows the formation of creatinine.

**Sulfur Metabolism.** The sulfur of the body is derived mainly from the sulfur-containing amino acids cystine, cysteine, and methionine. It is present in the body in such important constituents as insulin, glutathione, hair, nails, and taurocholic acid, a bile acid.

$$
\begin{array}{c}
\text{H} \quad\quad \text{OH} \\
| \quad\quad\quad | \\
\text{N} \sim \text{P} = \text{O} \\
\text{HN} = \text{C} \quad\quad \text{OH} \\
| \\
\text{N} - \text{CH}_2 - \text{COOH} \\
| \\
\text{CH}_3
\end{array}
\quad + \text{ ADP } \rightleftarrows
$$

Creatine phosphate

$$
\begin{array}{c}
\text{H} \\
| \\
\text{N} - \text{H} \\
\text{HN} = \text{C} \\
| \\
\text{N} - \text{CH}_2 - \text{COOH} \\
| \\
\text{CH}_3
\end{array}
\quad + \text{ ATP}
$$

Creatine

$$
\begin{array}{c}
\text{H} \\
| \\
\text{N} - \text{H} \\
\text{HN} = \text{C} \\
| \\
\text{N} - \text{CH}_2 - \text{COOH} \\
| \\
\text{CH}_3
\end{array}
\rightarrow
\begin{array}{c}
\text{H} \\
| \\
\text{N} \\
\text{HN} = \text{C} \quad \text{C} = \text{O} \\
| \\
\text{N} - \text{CH}_2 \\
| \\
\text{CH}_3
\end{array}
\quad + \text{ H}_2\text{O}
$$

Creatine                 Creatinine

The sulfur not used by the body is oxidized and eliminated in the urine as *inorganic sulfates* or as *ethereal sulfates* in which the sulfate is combined with organic radicals. A small fraction of unoxidized sulfur is also excreted, largely in the form of cystine. This is called *neutral sulfur*.

**Nucleoprotein Metabolism.** Nucleoproteins are composed of proteins conjugated with nucleic acids. During digestion the protein is split from the nucleic acids and is hydrolyzed to amino acids.

The nucleic acid is hydrolyzed to phosphoric acid, purines, pyrimidines, and a pentose sugar, ribose for RNA and deoxyribose for DNA. The phosphoric acid may be used for the synthesis of other phosphorus compounds or excreted in the urine. The pentose sugars follow the normal pathway of carbohydrate metabolism. The purines, adenine and guanine, are progressively oxidized by specific enzymes to uric acid and excreted in the urine.

The pyrimidines, cytosine, uracil, and thymine, are probably changed to ammonia and urea and excreted.

**Correlation of Carbohydrate, Lipid, and Protein Metabolism.** The correlation of the three major types of metabolism is shown in the diagram below. It will be noticed that pyruvic acid and acetyl coenzyme A are key compounds in the overall metabolism of food materials.

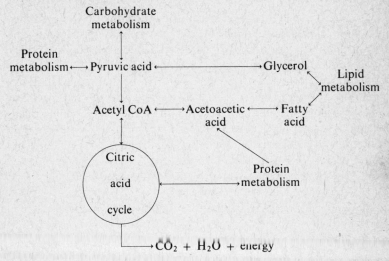

# 26. INORGANIC METABOLISM

Besides the three major classes of footstuffs, the body needs the mineral elements listed in Table 26.1. While these elements do not provide energy, they are necessary to metabolism and other body processes and must be supplied in the diet.

As shown in the table, about 96 per cent of the human body is composed of oxygen, carbon, hydrogen, and nitrogen which make up the water and organic constituents of the body. The other elements are present in the form of inorganic salts. The general functions of inorganic salts, briefly stated and later explained in more detail, are these:

1. They maintain the rigid structure of the body.
2. They build and repair tissues.
3. They maintain the normal contractility of muscles and the irritability of nerves.
4. They act as buffers and help maintain body neutrality.
5. They maintain a constant osmotic pressure.
6. They supply material for the production of digestive juices.
7. They act as cofactors for certain enzymes.

TABLE 26.1. ELEMENTARY COMPOSITION OF THE HUMAN BODY*

| Element | Per Cent | Approximate Amount (grams) in a 70-Kg. Man |
|---------|----------|-------------------------|
| Oxygen | 65.0 | 45,500 |
| Carbon | 18.0 | 12,600 |
| Hydrogen | 10.0 | 7,000 |
| Nitrogen | 3.0 | 2,100 |
| Calcium | 1.5 | 1,050 |
| Phosphorus | 1.0 | 700 |
| Potassium | 0.35 | 245 |
| Sulfur | 0.25 | 175 |
| Sodium | 0.15 | 105 |
| Chlorine | 0.15 | 105 |
| Magnesium | 0.05 | 35 |
| Iron | 0.004 | 3 |
| Manganese | 0.0003 | 0.2 |
| Copper | 0.0002 | 0.1 |
| Iodine | 0.00004 | 0.03 |

*From *Practical Physiological Chemistry* by Hawk, Oser, and Summerson, 13th ed. Copyright 1954, McGraw-Hill Book Company, Inc. Used by permission of McGraw-Hill Book Company.

# ELEMENTS OF SUBSTANTIAL QUANTITIES

The mineral elements that the body needs in substantial quantities are sodium, potassium, calcium, magnesium, chlorine, phosphorus, sulfur, and iron.

**Sodium.** Sodium is present in the animal body largely as sodium chloride and other inorganic salts. The sodium ion, which occurs principally in the extracellular tissue fluids, is essential in maintaining the normal osmotic pressure of the body fluids and the acid-base balance of the body. It acts as a stimulant in promoting the irritability of nerves and the movement of muscles. It also activates certain enzymes (e.g., apyrase) and inhibits others (e.g., phosphotransacetylase). Sodium bicarbonate plays an important role in the transportation of carbon dioxide from the cells to the lungs.

The sodium ion is rapidly absorbed from the intestinal tract. Most of the dietary sodium is excreted in the urine; small amounts are eliminated in the feces and sweat.

**Potassium.** The potassium ion occurs in all tissues, principally in intracellular fluids. Like the sodium ion, it promotes nerve and muscle irritability and assists in maintaining osmotic pressure. As an important constituent of red cells, the potassium ion participates in the carbon dioxide-carrying function of the blood. It also helps to maintain blood alkalinity. In addition, it activates certain enzymes required in carbohydrate metabolism, exerts a relaxing effect on the heart muscle between beats, and stimulates the excretion of water by the kidneys.

The potassium ion is rapidly absorbed from the intestinal tract. Most potassium excretion takes place by way of the urine.

**Calcium.** The calcium ion is necessary for the formation of bones and teeth, which are mainly hydroxyapatite, $Ca_3(PO_4)_2 \cdot Ca(OH)_2$. It is essential for the clotting of blood and the precipitation of milk casein in the stomach. It also helps to maintain the proper rhythm of the heartbeat.

The level of calcium in the blood is regulated by the parathyroid gland. Hypoactivity results in convulsions and tetany, hyperactivity in the hardening of organs such as the heart, lung, and arteries. Rickets and osteomalacia may develop when the body cannot absorb sufficient calcium. The presence of vitamin D is necessary for proper absorption of calcium and its deposition in the bones and teeth.

Calcium salts are excreted in both the feces and the urine. The daily requirement of calcium is 1 to 1.5 grams, pregnant and lactating females requiring a greater amount.

**Magnesium.** Magnesium compounds make up about one percent of bone tissue. The magnesium ion has a sedative effect on nerves and muscles. Injection of magnesium salts into the bloodstream produces anesthesia. Magnesium activates a number of enzymes required in the metabolism of carbohydrates involving ATP. Because magnesium compounds are not readily absorbed through the intestinal wall, they have been used as purgatives (Epsom salts). Deficiency of magnesium in the diet has produced convulsions and death.

**Chlorine.** The chloride ion is highly soluble in water and diffuses rapidly through the cell membrane. It assists in maintaining a normal electrolyte balance in the tissues and the normal osmotic pressure of the body fluids. It is also used by the body in the production of the hydrochloric acid of gastric juice.

**Phosphorus.** Inorganic phosphate combines with calcium and magnesium to form bones and teeth. It forms phospholipids, phosphoproteins, and nucleic acids. Phosphorus plays an important role in various phosphorylations essential for the metabolism of carbohydrates, such as the formation of hexose phosphates. Another important function is that it provides energy in forming the high-energy phosphate bond in phosphocreatine and adenosine phosphates. Phosphate salts act as buffers in the blood and urine.

Phosphorus is excreted in the urine in the form of inorganic phosphates. About one-third of phosphorus excretion is by way of the feces. The daily requirement of phosphorus is 1.32 grams.

**Sulfur.** Sulfur is derived mainly from methionine and cystine of the dietary protein. It is a constituent of body proteins, especially of the sulfur-rich protein keratin present in hair, nails, and skin. It also forms bile salts, insulin, and certain enzymes and coenzymes. Some toxic substances are detoxified by being combined with sulfate radicals and eliminated through the kidneys.

**Iron.** Iron-containing hemoglobin serves as a carrier of oxygen from the lungs to the tissues. Iron is also a component of myoglobin, a muscle pigment that combines with oxygen. It readily accepts and gives up electrons and is a necessary constituent of the electron transport and respiratory enzyme systems.

Iron is absorbed from the intestine in the form of the ferrous ion. That not immediately used by the body for hemoglobin or for other purposes is excreted in the feces or stored as *ferritin* in the liver, spleen, and bone marrow. Iron is carried in the plasma to the tissues in the form of an iron-protein called *transferrin*. The daily requirement of iron is 12 to 18 mg.

## TRACE ELEMENTS

The mineral elements which the body needs in trace amounts include copper, iodine, fluorine, cobalt, and zinc.

**Copper.** Copper is a catalyst for the formation of hemoglobin and the absorption and utilization of iron. It is a constituent of tyrosinase, an enzyme for the synthesis of melanin, and of certain respiratory enzymes. Most of the dietary copper is eliminated in the feces. The daily requirement of copper is 5 to 10 mg. In large quantities it is toxic.

**Iodine.** Iodine is a constituent of thyroxine, the chief hormone of the thyroid gland. Deficiency of iodine in the diet causes hypothyroidism, a common form being goiter. The daily requirement of iodine is 0.1 to 0.3 mg.

**Fluorine.** Most of the fluorine in the body is present in the bones and teeth, especially in the enamel of the teeth. One part per million of fluoride in drinking water has been found to reduce the incidence of tooth decay in children. Over 3 to 5 parts per million results in mottled enamel.

**Cobalt.** Cobalt is an essential constituent of vitamin $B_{12}$, which is concerned with formation of hemoglobin. Vitamin $B_{12}$ is used in treating pernicious anemia.

**Zinc.** Zinc is a constituent of insulin and of the enzyme carbonic anhydrase, which is important in respiration.

## WATER

Water is the most abundant compound in the human body, making up as much as 70 percent of the total body weight. It is of major importance to the normal functioning of the body.

1. It serves as a transport medium for food and for waste products.

2. It is necessary for the digestion (hydrolysis) of food.

3. It regulates body temperature through evaporation by way of the lungs and skin.

4. It serves as a medium for the biochemical reactions of the body.

5. It acts as a lubricant by keeping the tissues and joints moist.

6. It helps to maintain electrolyte balance by the process of osmosis.

Water is taken into the body in ingested fluids and in foods. It is also formed in the tissues by oxidation. Water is excreted from the lungs (expired air), skin (perspiration and sweat), intestines (feces), and kidneys (urine). The relative significance of these factors is shown in Table 26.2.

TABLE 26.2.   WATER BALANCE OF A NORMAL INDIVIDUAL

| Intake | | | | Output | |
|---|---|---|---|---|---|
| | Ml. | Percent of Total | | Ml. | Percent of Total |
| Drink | 1200 | 48 | Lungs | 400 | 16 |
| Water in food | 1000 | 40 | Skin | 500 | 20 |
| Metabolic water | 300 | 12 | Urine | 1400 | 56 |
| | | | Feces | 200 | 8 |
| Totals | 2500 | 100 | | 2500 | 100 |

In the normal individual the daily intake and output of water are in a state of equilibrium, and the individual is said to be in *water balance*. A water intake in excess of the ability of the body to eliminate water may be toxic (water intoxication). A water loss which exceeds water intake (dehydration) by 10 percent may result in illness and by 20 percent may lead to death.

# 27. BLOOD

The blood is the transport system of the body. It carries digested food material from the intestines, and oxygen from the lungs, to the tissues. It carries waste products, such as carbon dioxide, urea, and uric acid, to the organs of excretion (lungs, kidneys, intestine, and skin). It carries hormones, enzymes, and vitamins to different parts of the body where they are to function.

Blood also distributes heat and helps to maintain body temperature. It aids in the maintenance of acid-base balance and water balance. Through the action of the white corpuscles, antibodies, and antitoxins, the blood protects the body from pathological microorganisms.

## COMPOSITION OF THE BLOOD

Blood is a viscous, slightly alkaline fluid. It consists of formed elements, or cells, suspended in the liquid plasma. Some important chemical constituents of the blood, their normal concentration and their variation in certain pathological conditions, are given in Table 27.1.

**Formed Elements.** The formed elements, which constitute 45 percent of the blood by volume, include the erythrocytes (red cells or corpuscles), the leukocytes (white cells), and the thrombocytes (platelets).

ERYTHROCYTES. There are about 5,000,000 red blood cells per cubic millimeter of blood. Their main function is to carry oxygen from the lungs to the cells. They also remove carbon dioxide from the cells and carry it to the lungs.

The red blood cells are formed in the bone marrow. Their life span is about four months. Old cells are broken down in the spleen, releasing bile pigments and iron.

*Hemoglobin.* Hemoglobin, the respiratory pigment of the blood that gives the color to the red cells, has a molecular weight of about 67,000. The hemoglobin molecule is made up of four

TABLE 27.1.   CONCENTRATION OF BLOOD CONSTITUENTS
IN HEALTH AND DISEASE

| Constituent | Average Value* per 100 Ml. of Blood | Pathological Conditions | |
|---|---|---|---|
| | | Increased in | Decreased in |
| Hemoglobin, male (B)† | 15.9 g. | Polycythemia | Protein deficiency; anemia |
| Total protein (S) | 7.2 g. | Multiple myeloma; anhydremia | Nephrosis; protein deficiency; liver disease |
| Albumin (S) | 5.2 g. | Anhydremia | Nephrosis; protein deficiency; liver disease |
| Globulin (S) | 2.0 g. | Multiple myeloma | |
| Nonprotein nitrogen (B) | 29 mg. | Nephritis | |
| Urea nitrogen (B) | 13.6 mg. | Nephritis | Nephrosis |
| Creatinine (P) | 1.0 mg. | Nephritis | |
| Uric acid (S) | 4.4 mg. | Gout, arthritis | |
| Glucose (B) | 90 mg. | Diabetes mellitus | Hyperinsulinism |
| $CO_2$ capacity (P) | 60% by vol. | Alkalosis | Diabetic acidosis |
| Cholesterol, total (S) | 210 mg. | Nephrosis | Hyperthyroidism |
| Calcium (S) | 10.0 mg. | Hyperparathryoidism | Hypoparathyroidism |
| Phosphorus (S) | 3.6 mg. | Severe nephritis | Rickets |
| NaCl (P) | 595 mg. | Nephritis; eclampsia | Pneumonia; prolonged vomiting |
| Bilirubin (S) | 0.54 mg. | Obstructive jaundice; hemolytic anemia | Secondary anemia |

*Most of these values were taken from *Biochemistry of Disease* by Bodansky and Bodansky, 2nd ed. Copyright 1952 by The Macmillan Company.
† B = whole blood; S = serum; P = plasma.

molecules of the iron-containing heme conjugated with one molecule of the protein globin (a histone).

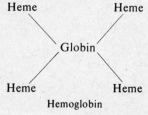

Hemoglobin

Heme consists of four pyrrole rings joined by —CH groups (porphyrin) and an atom of iron in the ferrous state (see p. 167).

*Anemia.*   Anemia is a condition in which the number of red cells or the amount of hemoglobin is below normal.   It may result from any one of the following:

1. Hemorrhage.

2. Deficiency of iron (iron deficiency anemia or nutritional anemia).

3. Deficiency of folic acid (macrocytic anemia of infancy and pregnancy).

4. Lack of the intrinsic factor or deficiency of vitamin $B_{12}$, resulting in lowered activity of the blood-producing tissues (pernicious anemia).

5. Abnormal destruction of red cells (hemolytic anemia).

6. Destruction of bone marrow by chemicals, drugs, or radiation (aplastic anemia).

7. Abnormal hemoglobins such as those causing sickle-cell anemia and thalassemia.

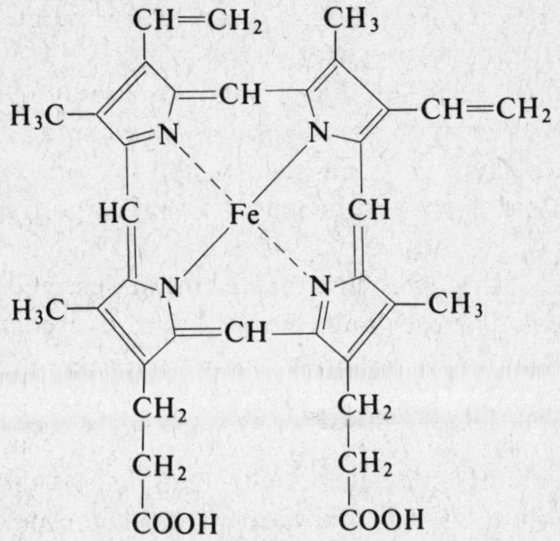

Heme

LEUKOCYTES. The white cells are larger than the red cells and have a nucleus. There are about 6,000 white cells per cubic millimeter of blood. They act as phagocytes, engulfing and destroying harmful microorganisms. The white cell count rises sharply in acute infections. In leukemia there is an excessive increase in their number.

THROMBOCYTES. The thrombocytes, or platelets, are irregularly shaped bodies without nuclei. They are smaller than red cells and number about 250,000 per cubic millimeter of blood. They are concerned with the coagulation of blood and contain cephalin, a phospholipid involved in the early stages of clotting.

*Coagulation of Blood.* The generally accepted theory of the blood-clotting mechanism is as follows:

1. The enzyme *thrombin* is formed from its inactive form, *prothrombin*, by the action of thromboplastin and calcium ions. *Thromboplastin* is a cephalin-protein substance released from the thrombocytes and injured tissue cells when blood is shed.

$$\text{Prothrombin} \xrightarrow[\text{Ca}^{++}]{\text{thromboplastin}} \text{Thrombin}$$

2. The thrombin, by enzyme action, converts *fibrinogen*, a soluble plasma protein, into *fibrin*, the clot.

$$\text{Fibrinogen} \xrightarrow{\text{thrombin}} \text{Fibrin}$$

*Serum.*  When blood clots, the formed elements or corpuscles are enmeshed in the threadlike mass of the insoluble fibrin. The clot, on standing, exudes serum, a clear yellow liquid. Serum is thus plasma without fibrinogen.

*Anticoagulants.*  The anticoagulants are useful in preventing blood from coagulating inside the blood vessels and in preserving blood for purposes of blood transfusion or laboratory tests. *Heparin* is a natural antiprothrombin found largely in the liver. *Dicumarol*, a popular anticoagulant drug, also has antiprothrombin properties. *Hirudin* is an antithrombin secreted by bloodsucking leeches. *Oxalates* and *citrates* are also used. They prevent blood from clotting by precipitating the calcium ions or repressing their ionization, respectively.

*Hemophilia.*  Hemophilia is a hereditary disease in which the blood shows considerable delay in clotting. This is thought to be due to the absence or deficiency of an antihemophilic globulin which is necessary to thrombin formation.

**Plasma.**  The plasma is a clear fluid constituting about 55 per cent of the blood by volume.

PLASMA PROTEINS.  The most important protein components of the plasma are the serum albumins, the serum globulins, and fibrinogen. *Serum albumin* is essential in maintaining normal osmotic pressure and therefore the water balance of the body. A deficiency of serum albumin causes excessive passage of water from the blood vessels into the surrounding tissues, producing edema. The *serum globulins* consist of alpha, beta, and gamma globulins. Gamma globulin contains most of the antibodies and antitoxins the body uses to combat infection. *Fibrinogen* is concerned with the clotting of the blood.

OTHER SUBSTANCES IN THE PLASMA.  The plasma contains

nutrients (glucose, amino acids, and lipids); nitrogenous waste products (urea, uric acid, creatinine, and creatine); lactic acid; bile pigments; and a small amount of acetone bodies. It also contains hormones, vitamins, enzymes, and antibodies. The electrolytes present in the plasma include $Na^+$, $K^+$, $Ca^{+2}$, $Mg^{+2}$, $Cl^-$, $HCO_3^-$, $HPO_4^{-2}$, and $H_2PO_4^-$.

## RESPIRATION

Respiration is the process by which the body, with the aid of the blood, takes in oxygen and gives off carbon dioxide. The principal chemical reactions involved are presented below.

**Transport of Oxygen to the Tissues.** Because of the higher partial pressure of oxygen in the lungs (100 mm. Hg), it diffuses into the plasma and enters the red cells where it combines with hemoglobin (HHb)* to form oxyhemoglobin ($HHbO_2$).

$$O_2 + HHb \rightarrow HHbO_2$$

The oxyhemoglobin, which is acid in reaction, reacts with bicarbonates to form carbonic acid and a basic form of oxyhemoglobin ($KHbO_2$). The carbonic acid, in the presence of the enzyme carbonic anhydrase, decomposes to carbon dioxide and water.

$$HHbO_2 + KHCO_3 \rightleftharpoons H_2CO_3 + KHbO_2$$

$$H_2CO_3 \xrightarrow[\text{anhydrase}]{\text{carbonic}} CO_2 + H_2O$$

The basic oxyhemoglobin is carried to the tissues by the blood. Because of the low partial pressure of oxygen in the tissues (40 mm. Hg), the basic oxyhemoglobin decomposes to oxygen and basic hemoglobin.

$$KHbO_2 \rightarrow O_2 + KHb$$

The oxygen released diffuses into the tissue cells where it enters into metabolic reactions. The reduced hemoglobin returns to the lungs in the venous blood.

**Transport of Carbon Dioxide to the Lungs.** The carbon dioxide formed by metabolic reactions in the tissues diffuses into the plasma and enters the red cells where the enzyme carbonic anhydrase, now working in reverse, catalyzes its combination with water to form carbonic acid.

*Hemoglobin is written here as HHb to indicate its deoxygenated form.

$$CO_2 + H_2O \xrightarrow[\text{anhydrase}]{\text{carbonic}} H_2CO_3$$

The carbonic acid reacts with basic hemoglobin to form bicarbonates and acid-reacting hemoglobin.

$$H_2CO_3 + KHb \rightarrow KHCO_3 + HHb$$

As the bicarbonate ion concentration increases, these ions diffuse out of the red cells and into the plasma. To balance the loss of negative ions from the red cells, an equal number of chloride ions enter the red cells from the plasma. This is called the *chloride shift*.

Some of the carbon dioxide that diffuses into the red cells combines with the amino group ($NH_2$) of hemoglobin to form carbaminohemoglobin (HHbNHCOOH).

$$HHbNH_2 + CO_2 \rightarrow HHbNHCOOH$$

Carbon dioxide, mainly as bicarbonates but partly as carbonic acid and carbaminohemoglobin, is returned to the lungs. There the carbon dioxide is released from the carbaminohemoglobin and excreted. The reduced hemoglobin is oxygenated in the lungs; the acidic oxyhemoglobin thus formed combines with the bicarbonates to form carbonic acid which releases carbon dioxide for excretion. The bicarbonate ions in the plasma diffuse into the red cells to replace the bicarbonate that was converted to carbon dioxide and in turn release carbon dioxide by the above process. A reverse chloride shift takes place to balance the loss of negative ions from the plasma.

The various reactions taking place in the lungs and in the tissues are diagrammed in Fig. 27.1.

## ACID-BASE BALANCE

The normal pH of the blood is 7.4. Acid-base balance is maintained in the following ways:

*1. By the action of the buffer systems of the blood.* A buffer system is a combination of a weak acid and its salt or a weak base and its salt which can tolerate additions of acid or alkali without significant changes in pH. The most important buffer system of the blood is the carbonic acid-bicarbonate system. To maintain a pH in the blood of 7.4, the ratio of carbonic acid to bicarbonate must be 1 to 20. Acids entering the blood are neutralized by the

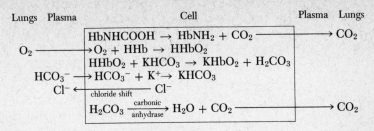

In the Lungs

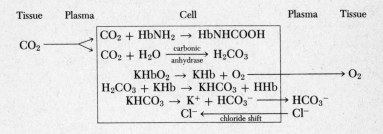

In the Tissues

FIG. 27.1. Respiration reactions in the lungs and tissues. From Routh: *20th Century Chemistry*, 3rd ed. Philadelphia, W. B. Saunders Co., 1963.

bicarbonate and alkalies by the carbonic acid. The phosphate buffer system is the combination of an acid salt and a basic salt. Hemoglobin and its salts and proteins and their alkali salts also play a part in the maintenance of blood pH. These buffer systems are illustrated below.

| $\dfrac{H_2CO_3}{NaHCO_3}$ | $\dfrac{NaH_2PO_4}{Na_2HPO_4}$ | $\dfrac{HHb}{NaHb}$ | $\dfrac{HHbO_2}{NaHbO_2}$ | $\dfrac{H\,Protein}{NaProtein}$ |
|---|---|---|---|---|

*2. By the retention or excretion of carbon dioxide by the lungs, involving the carbonic acid-bicarbonate buffer system.* When there is an excess of carbonic acid in the blood, respiration is stimulated and carbonic acid is decomposed to carbon dioxide, which is expelled from the lungs, bringing the ratio of carbonic acid to bicarbonate back to the normal 1 to 20. Conversely, when there is an excess of bicarbonate in the blood, respiration is decreased and therefore the carbonic acid concentration increased.

*3. By the production of ammonia by the kidneys.* This ammonia forms ammonium salts with the acid products of protein metabolism. The ammonium salts are excreted in the urine, thus conserving sodium and potassium for the alkali reserve of the blood.

4. *By the excretion of acidic or basic phosphate by the kidneys.*

**Acidosis and Alkalosis.**    Disturbances in the acid-base balance result in acidosis or alkalosis.  The amount of bicarbonate stored in the blood which can be used to neutralize acids is called the *alkali reserve.*  When most of this is used up the body cannot protect itself against acids.  The result is *acidosis.*  Acidosis is also caused by an abnormal retention or an excess of acid.  *Alkalosis*, conversely, is caused by a large increase in alkali reserve or a significant loss or excretion of acid.

Acidosis and alkalosis can be metabolic or respiratory in cause. They can also be compensated or uncompensated.  In *metabolic acidosis* there is an abnormal decrease of blood bicarbonate caused by increased formation of acetone bodies (diabetes mellitus, ketosis), excessive ingestion of acid-forming foods, or diarrhea.  *Respiratory acidosis* is an abnormal increase of blood carbonic acid caused by hypoventilation resulting from morphine poisoning, pneumonia, or emphysema.  Acidosis can be *compensated* by increased pulmonary ventilation, increased ammonia formation in the kidneys, or increased acid excretion by the kidneys.  The pH of the blood is maintained in its normal range.  In *uncompensated acidosis*, the pH is significantly lowered.

*Metabolic alkalosis* is an abnormal increase in blood bicarbonate caused by overdosage of sodium bicarbonate or by excessive loss of hydrochloric acid from persistent vomiting or gastric lavage.  In *respiratory alkalosis* there is an abnormal decrease of blood carbonic acid caused by hyperventilation due to fever, high altitude, hot baths, or extreme anxiety.  Alkalosis can be *compensated* by increased excretion of bicarbonate or alkaline phosphate in the urine and by decreased ammonia formation in the kidneys. When it is *uncompensated* the pH of the blood is significantly raised.

# 28. URINE

The waste products of the body are eliminated by four organs: the skin, the lungs, the intestines, and the kidneys. The main task of excretion is carried out by the kidneys. Inorganic salts, water, and nitrogenous waste material are excreted through the urine. By the filtration and reabsorption processes, the kidneys determine the quantities of substances to be excreted or retained and therefore maintain at normal levels the pH, osmotic pressure, volume, and composition of the blood and other body fluids.

**Formation of Urine.** The kidney is made up of a million or more filtration units called *nephrons* or *malpighian corpuscles*. Each nephron consists of a bundle of capillaries called a *glomerulus* surrounded by a capsule known as *Bowman's capsule*. Bowman's capsule elongates into a tortuous tubule whose distal end enters a collecting duct leading to the pelvis of the kidney. (See Fig. 28.1.)

Urine is formed in the kidneys by two distinct processes: glomerular filtration and tubular reabsorption.

GLOMERULAR FILTRATION. As the blood passes through the glomerulus the constituents other than blood cells and protein filter through the capillary walls and enter the tubules.

TUBULAR REABSORPTION. During the passage of the filtrate along the tubules, a large portion of water and substances of value to the body such as glucose, certain inorganic salts, and amino acids, are reabsorbed into the blood. About 99 per cent of the 170 liters of filtrate that pass through the glomeruli each day is reabsorbed, leaving only about 1.5 liters to be excreted as urine.

**Volume of Urine.** From 1,000 to 1,500 ml. of urine are excreted per day by the normal adult. The volume depends on the amount of liquid consumed, the amount of water excreted by the skin, and the amount lost during such conditions as fever or diarrhea. *Polyuria*, or above normal secretion of urine, occurs in certain diseases such as diabetes mellitus and as a result of the intake of diuretics such as caffeine.

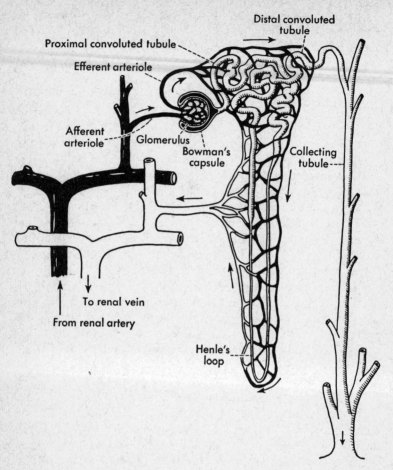

FIG. 28.1. The microscopic anatomy of a single kidney tubule, showing its blood supply. From Hunter and Hunter: *College Zoology*. Philadelphia, W. B. Saunders Co., 1949.

**Physical Properties of Urine.**   The color, odor, specific gravity, and pH of urine are significant qualities in urine analysis.

COLOR.   Normal urine is pale yellow to amber in color, depending on the amount excreted.   The chief pigment in urine is urochrome.   Because the daily excretion of urochrome is fairly constant, the larger the volume of urine voided, the paler the color.   The color of urine is also affected by the presence of certain drugs, dietary substances, and abnormal constituents in disease conditions.

Normal urine is clear. It may be cloudy after a heavy meal and it becomes cloudy on standing because of the precipitation of phosphates.

ODOR. Fresh urine has a faintly aromatic odor. It becomes ammoniacal on standing. The ingestion of certain foods and drugs and the presence of abnormal constituents in disease may affect the odor.

SPECIFIC GRAVITY. The specific gravity of normal urine ranges from 1.008 to 1.030. It varies according to the concentration of solid waste products; the greater the volume excreted, the lower the specific gravity.

pH. Normal urine is usually slightly acidic, with a pH of 5.5 to 7.5, because of the presence of acid phosphates and sulfates and organic acids. A high protein meal will increase the acidity; a fruit and vegetable diet will produce an alkaline urine because of the alkaline residue formed in metabolism. The urine usually becomes temporarily alkaline after a meal. An equivalent of alkaline salts enters the blood for each equivalent of hydrochloric acid secreted into the stomach during digestion, and these salts are excreted in the urine. This temporary shift in pH is called the *alkaline tide*. An acidic urine is produced in diabetes mellitus and in starvation because of the excretion of ketone bodies.

**Normal Constituents of the Urine.** Normal urine consists of about 96 percent of water and 4 percent of dissolved organic and inorganic waste products. The main constituents of normal urine and the average quantity excreted in twenty-four hours are given in Table 28.1.

UREA. Urea is formed in the liver from the oxidative deamination of amino acids. It is the chief nitrogenous end product of protein metabolism. The amount of urea in the urine varies directly with the protein ingested and usually represents from 80 to 90 percent of the total nitrogen excreted. Urea excretion is increased in fever and diabetes and decreased in diseases of the liver and kidney.

URIC ACID. Uric acid is the chief end product of the metabolism of purines, which are derived from nucleic acids and, in turn, nucleoproteins. It is usually excreted in the form of salts of uric acid. The inclusion in the diet of foods rich in nucleoproteins, such as liver, kidney, sweetbreads, and heart, increases the uric acid output in the urine. The excretion of uric acid is increased in leukemia and decreased in gout and nephritis. Under certain conditions uric acid crystallizes in the kidneys, forming kidney stones.

CREATININE.   Creatinine, the anhydride of creatine, is thought to originate mainly from the creatine of the muscles. The amount of creatinine excreted is nearly constant for an individual (1 to 1.7 grams per day) and is relatively independent of the protein intake. Rather, it is proportional to the muscular development of the individual. The milligrams of creatinine excreted per day per kilogram of body weight is called the *creatinine coefficient*.

### TABLE 28.1.   NORMAL CONSTITUENTS OF THE URINE

| | Grams per 24 Hours |
|---|---|
| Water | 1200.0 |
| Total solids | 60.0 |
| Organic: | |
|    Urea | 30.0 |
|    Uric acid | 0.7 |
|    Creatinine | 1.4 |
| Inorganic: | |
|    Chlorides (as NaCl) | 12.0 |
|    Phosphates (as $P_2O_5$) | 2.5 |
|    Sulfates (as $SO_3$) | 2.0 |
|    Ammonia | 0.7 |

CHLORIDES.   The chief chloride in the urine is sodium chloride. The amount of this chloride excreted per day is usually from 10 to 15 grams, though the amount varies with the intake of salt in the diet. The urinary excretion of sodium chloride is decreased in fevers, pneumonia, and nephritis. Retention of sodium chloride in the blood and tissues due to nephritis causes water to accumulate in the tissues, resulting in edema.

PHOSPHATES.   The phosphates in the urine are derived from phosphorus-containing foods such as casein, nucleoproteins, and phospholipids. About 1.8 grams of phosphates are excreted daily. Phosphate excretion is increased in rickets, osteomalacia, and pulmonary tuberculosis, and decreased in acute infectious diseases, kidney disease, and the later stages of pregnancy during fetal bone formation.

SULFATES.   More than three-fourths of the total sulfur excreted is in the form of inorganic sulfates. The sulfates in the urine are derived mainly from the oxidation of the sulfur-containing amino acids cystine, cysteine, and methionine in the dietary protein. The urinary sulfate excretion is increased in acute fevers and tissue breakdown, and decreased in loss of appetite and impaired renal function.

AMMONIA.   The ammonia in the urine is usually in the form of ammonium chloride, ammonium sulfate, and ammonium phosphate.   An increase in excretion is caused by acid-forming foods such as meat and cereals, and a decrease by alkali-forming foods such as fruits and vegetables.   Excretion is increased in diabetic acidosis.

OTHER CONSTITUENTS.   Small amounts of hippuric acid, indican, amino acids, and urochrome may also be present in normal urine.

**Abnormal Constituents of the Urine.**   The appearance of certain substances in the urine usually indicates a pathological condition. The most common abnormal constituents are glucose, proteins, ketone bodies, bile, blood, and large amounts of indican.

GLUCOSE.   Normally there is a very small amount of glucose in the urine, not enough to give a positive Fehling's or Benedict's test.   Large amounts of glucose in the urine (glycosuria) may be due to several causes: a lack of insulin (*diabetes mellitus*), a lowered renal threshold (*renal glycosuria*), or a high carbohydrate meal (*temporary glycosuria*).

Lactose is sometimes present in the urine of lactating females (*lactosuria*).   The pentose xylulose occurs in the urine (*pentosuria*) of some individuals.

PROTEINS.   Owing to their large size, protein molecules are incapable of filtering through the kidney.   Their presence in the urine (*proteinuria* or *albuminuria*) usually is an indication of impaired kidney function such as occurs in nephritis and nephrosis.

KETONE BODIES.   The excretion of acetone or ketone bodies in the urine (*acetonuria* or *ketonuria*) is increased in uncontrolled diabetes, high fevers, and starvation.   Two of the three ketone bodies are acids (acetoacetic acid and $\beta$-hydroxybutyric acid) and must be neutralized prior to excretion.   Therefore the presence of ketone bodies in any considerable amount in the urine is indicative of *acidosis*, the result of the depletion of the body's alkali reserve.

BILE.   Bile is normally excreted in the feces. Its presence in the urine, shown by the yellowish green to brown color and by yellow foam when shaken, indicates obstruction in the bile duct (*obstructive jaundice*), excessive destruction of red blood cells (*hemolytic jaundice*), or severe liver damage.

BLOOD. Blood may appear in the urine as red blood cells (*hematuria*) or as hemoglobin (*hemoglobinuria*). Hematuria is caused by lesions either in the kidney or in the urinary tract. Hemoglobinuria is due to hemolysis of the red blood cells and occurs in such diseases as malaria, scurvy, smallpox, and scarlet fever.

Pus and casts caused by infection in the kidneys or the urinary tract may also be present in the urine.

INDICAN. When protein foods containing the amino acid tryptophan putrefy in the large intestine, indole and skatole are formed. These substances are detoxified in the liver, indole being converted to indican (indoxyl potassium sulfate), which is excreted in the urine. While small amounts of indican occur in normal urine, larger amounts usually indicate an excessive amount of putrefaction in the large intestine.

**Kidney Function Tests.** A number of tests have been devised to check on the function of the kidneys. Several of these are described below.

UREA CLEARANCE TEST. This test involves a comparison between the concentration of urea in the blood and the rate of its excretion in the urine. Normally 75 ml. of blood is cleared of urea in one minute. A urea clearance showing less than 65 percent of normal kidney function implies impairment.

PHENOLSULFONPHTHALEIN TEST (PSP). In this test phenolsulfonphthalein (phenol red) is injected either intravenously or intramuscularly. Normally functioning kidneys will excrete 25 percent or more of the dye in the urine during the first 15 minutes, and 60 to 80 percent after two hours. Less than 50 percent elimination within two hours implies decreased kidney function.

CONCENTRATION TEST. Properly functioning kidneys have the power of conserving water by a higher degree of reabsorption. Thus if fluids are withheld for 12 hours, normal urine will have a high specific gravity of 1.025 or above. Lower values indicate damaged kidneys.

# 29. VITAMINS

Vitamins are organic compounds other than carbohydrates, fats, and proteins that are essential to animal and human life. Not all species of animals require the same vitamins in their diet. Guinea pigs and the primates, including man, need vitamin C, while rats, dogs, and other species have the ability to synthesize it. In man, vitamin K and some of the B vitamins are synthesized in the intestine by bacterial action. Vitamin $D_3$ is formed from 7-dehydrocholesterol in the skin by irradiation from the ultraviolet rays in sunlight. Those vitamins that have precursor forms (provitamins) are ingested as such and converted in the body. For example, carotenes in fruits and vegetables are changed into vitamin A in the intestinal tract.

Although the amounts of vitamins needed by the body are small compared with the amounts required of carbohydrates, fats, and proteins, deficiency of these accessory food factors results in serious diseases specific for each vitamin. Because vitamins may be inactivated by heat, light, oxidation, and the presence of alkalies, care must be taken in the preparation and cooking of food.

Many of the vitamins, particularly those of the B complex, are part of coenzyme systems that play vital roles in metabolism. Some vitamins are concerned mainly with growth, reproduction, and proper utilization of mineral elements, while others help to maintain healthy tissues, nerve stability, and proper functioning of the digestive tract.

**Historical.** Vitamin deficiency diseases such as scurvy and beriberi were the lot of sailors for centuries because of the lack of fresh foods. It was found that scurvy could be prevented during long voyages by the addition of citrus fruits to the diet. In 1882 a Japanese medical officer named Takaki eradicated beriberi from the Japanese navy by giving the sailors increased quantities of fruit, vegetables, barley, and meat. However, beriberi was considered to be an infectious disease until 1897 when Eijkman, a Dutch physician, proved experimentally that it had a dietary cause. He produced polyneuritis in chickens by feeding them polished rice and then cured the disease by giving them the polish-ings. He concluded that the rice polishings contained a substance that neutralized the beriberi toxin. In 1911 a Polish chemist named Funk isolated thiamine from rice polishings and proved that it prevented and cured beriberi. He coined the word *vitamine* (vital amine) for these essential dietary substances. Later as more vitamins were discovered that were not amines, the spelling was changed to *vitamin*. Vitamin A was found in cod liver oil and butter by McCollum and Davis and Osborne and Mendel in 1913.

**Classification.** The vitamins may be classified according to solubility into two large groups: the fat-soluble vitamins and the water-soluble vitamins. The fat-soluble vitamins include vitamins A, D, E, and K. The water-soluble vitamins include the B com-plex and C.

## THE FAT-SOLUBLE VITAMINS

Vitamins A, D, and K are not appreciably affected by heat and therefore can withstand boiling without losing their potency. Vitamin E is heat-stable up to 200°C. in the absence of oxygen.

**Vitamin A.** Vitamin A is a complex, high molecular weight, primary alcohol, made up of a long unsaturated side-chain at-tached to a six-carbon ring, as shown by its formula below.

Vitamin A

Vitamin A is found in animal foods such as fish liver oil, milk, and butter. Precursors of this vitamin are present as carotenoid pigments in yellow vegetables such as carrots and sweet potatoes. The provitamins include $\alpha$-, $\beta$-, and $\gamma$-carotene and crypto-xanthin. When ingested into the animal body, they are converted into vitamin A.

Vitamin A is an important constituent of rhodopsin, or visual purple, which is essential for vision in dim light. This explains why a deficiency of this vitamin is a common cause of night blindness.

Vitamin A also promotes growth and aids in maintaining healthy epithelial cells of the mucous membranes of the eyes, skin, mouth tissue, the gastrointestinal tract, the respiratory tract, and the genitourinary system. Thus it protects the body against infection and keratinization, the drying up and the hardening of the mucous membrane. The common vitamin A deficiency disease *xerophthalmia* is caused by keratinization inside the eye.

**Vitamin D.** There are a number of forms of vitamin D, but vitamin $D_2$ and vitamin $D_3$ are the only ones of importance in nutrition. Both are related to the sterols and have similar structures.

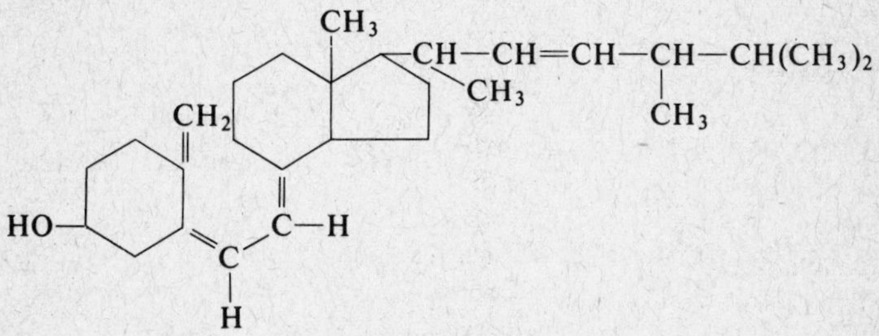

Vitamin $D_2$ (calciferol)

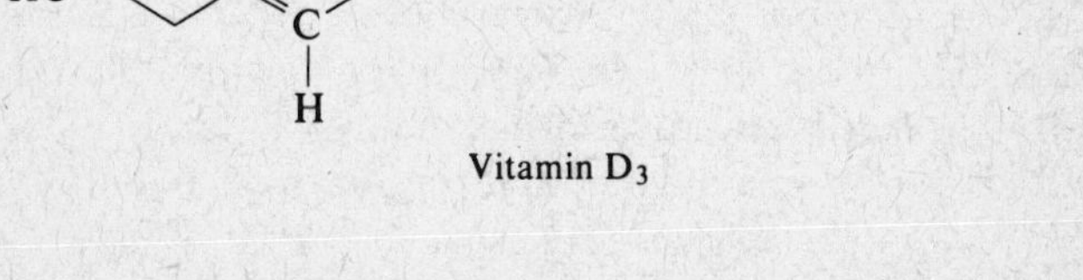

Vitamin $D_3$

Vitamin $D_2$, known as calciferol, is formed by the irradiation of the provitamin ergosterol, which is found in yeast and ergot. Vitamin $D_3$ is devived from 7-dehydrocholesterol present in the fatty tissue under the skin and is converted to vitamin $D_3$ by the ultraviolet rays of the sun. Both vitamins have the same physiological functions, and the term vitamin D refers to either one.

Vitamin D is essential to the proper absorption of calcium and phosphorus from the intestinal tract and their deposition in the bones and teeth. Deficiency of vitamin D produces rickets in children and osteomalacia in adults. Rickets may be prevented or remedied by giving foods rich in vitamin D, such as fish liver oil, egg yolk, and Viosterol (irradiated ergosterol), or by exposure to ultraviolet light. Overdosage of this vitamin may cause toxic symptoms due to deposition of calcium in the blood vessels and kidney tubules.

**Vitamin E.** Vitamin E is a high molecular weight alcohol and occurs in nature in three related forms, known as $\alpha$-, $\beta$-, and $\gamma$-tocopherol. Among these, $\alpha$-tocopherol is the most potent. Its structural formula is given below.

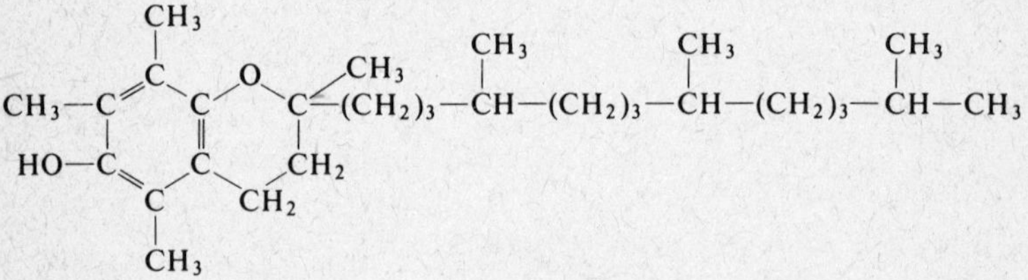

$\alpha$-Tocopherol

Because vitamin E has been found to be essential for normal reproduction in the rat, it has been called the *antisterility vitamin*. In addition to sterility, vitamin E deficiency may cause muscular dystrophy and paralysis in rats and other animals. However, the evidence that vitamin E is essential to man has not been established. Vitamin E is widely distributed in plant and animal tissues. The richest source is wheat germ oil.

**Vitamin K.** Vitamin K is a derivative of 1,4-naphthoquinone. It exists in at least two active forms, vitamin $K_1$ and vitamin $K_2$. The structural formula of vitamin $K_1$ is shown below.

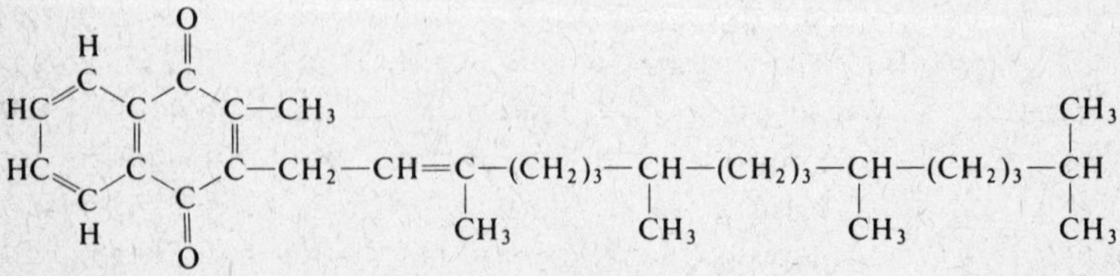

Vitamin $K_1$

Vitamin K is essential for the synthesis of prothrombin, one of the factors involved in the normal process of blood clotting. A deficiency of this vitamin causes a prolonged blood clotting time, which may result in excessive bleeding.

The best dietary sources of vitamin $K_1$ are green alfalfa and other green vegetables. Vitamin $K_2$ is produced by bacterial action in the intestines. Since vitamin K is fat-soluble, an adequate supply of bile is necessary for its absorption from the intestine. This is the reason why persons with obstructive jaundice and related diseases often have low prothrombin levels. For such persons, vitamin K and bile are usually adminstered before surgery to prevent profuse bleeding. Vitamin K has also been useful in checking bleeding in newborn infants, whose prothrombin level may be low.

## THE WATER-SOLUBLE VITAMINS

The water-soluble vitamins include the B complex vitamins and vitamin C. While vitamin C is a single vitamin, the vitamin B complex represents a large number of individual vitamins. These vitamins were formerly identified as vitamin $B_1$, $B_2$, G, etc. However, because this type of nomenclature resulted in much confusion, it is the common practice today to designate the B vitamins by their chemical names. The following is a list of the more important members of the vitamin B complex:

> Thiamine (Vitamin $B_1$)
> Riboflavin (Vitamin $B_2$)
> Nicotinic acid (Niacin)
> Pyridoxine (Vitamin $B_6$)
> Pantothenic acid
> Biotin
> Inositol
> Choline
> p-Aminobenzoic acid
> Folic acid
> Cyanocobalamin (Vitamin $B_{12}$)

The B vitamins are not chemically related. Their functions are quite varied; however, most of them are constituents of coenzymes. A number of them can be synthesized in the intestine by microorganisms.

Most of the B vitamins are affected by light. They are fairly stable to the moderate heat of cooking. Vitamin C is the least stable of the vitamins. Even short periods of cooking tend to destroy its potency unless the vegetable is initially placed in boiling water from which the oxygen has passed off.

**Thiamine (Vitamin $B_1$).** Thiamine is a sulfur-containing amine. It consists of a pyrimidine ring and a thiazole ring joined by a methylene ($CH_2$) group as shown below.

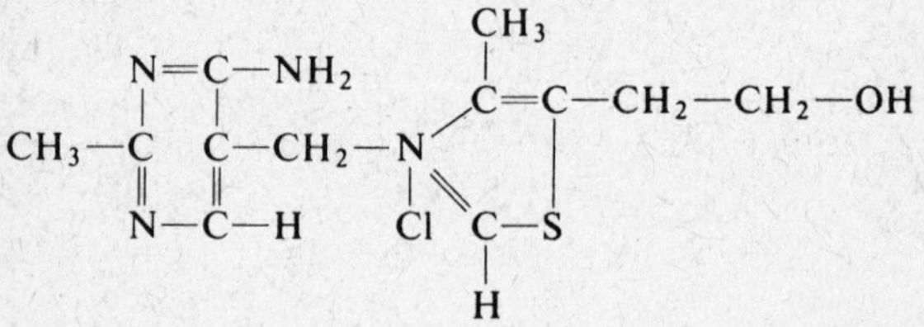

Thiamine

Thiamine in the tissues combines with two molecules of phosphoric acid to form the coenzyme thiamine pyrophosphate (cocarboxylase). This coenzyme is required by pyruvic oxidase in splitting carbon dioxide from pyruvic acid, a product of the intermediate metabolism of carbohydrates.

Besides being essential in the utilization of carbohydrates by the body, thiamine is necessary for normal growth, for promoting appetite, for digestion, for tonicity of the digestive tract, and for the normal functioning of the nervous and cardiovascular systems.

Deficiency of thiamine results in beriberi, called polyneuritis in animals. Early symptoms include loss of appetite, fatigue, weakness, constipation, and muscular cramps. Later symptoms include muscular incoordination, edema, paralysis, atrophy of muscle, and finally death as a result of heart failure.

Diseases due to thiamine deficiency may be cured by giving foods rich in thiamine such as rice polishings, brewer's yeast, and wheat germ, or pure thiamine.

**Riboflavin (Vitamin $B_2$ or G).** Riboflavin is an orange-yellow, needle-like crystalline solid. It is composed of an isoalloxazine nucleus combined with reduced ribose. The structure of riboflavin is as follows:

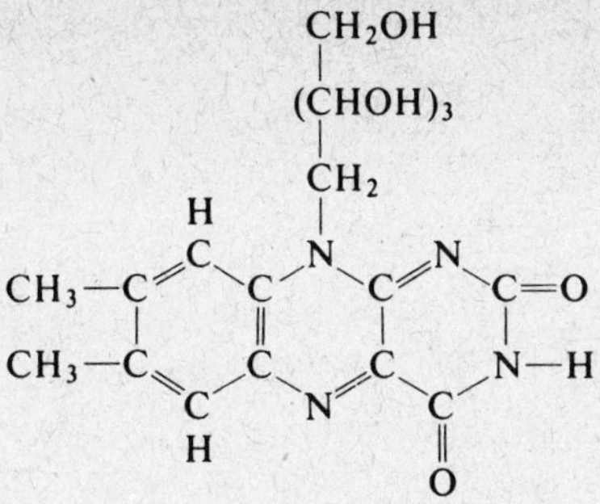

$$CH_2OH$$
$$(CHOH)_3$$
$$CH_2$$

Riboflavin

Riboflavin is essential for oxidation reactions in the cells. It is converted in the body to flavin mononucleotide (FMN) and flavin adenine dinucleotide (FAD), coenzymes for certain dehydrogenases.

A deficiency of riboflavin in the diet results in characteristic lesions on the tongue, lips, and in the corners of the mouth, and oily skin eruptions in the folds about the nose. Other symptoms may also appear, such as ocular disturbances with burning and itching, lacrimation, conjunctivitis, and corneal opacity.

Dietary sources of riboflavin include brewer's yeast, liver, milk, eggs, and leafy vegetables.

**Nicotinic Acid.** Nicotinic acid is a derivative of pyridine. It is a degradation product of nicotine, the alkaloid of tobacco, but it is not formed from tobacco in the body. To avoid association with nicotine, nicotinic acid is often called *niacin*. The structure of nicotinic acid and of its amide are given below.

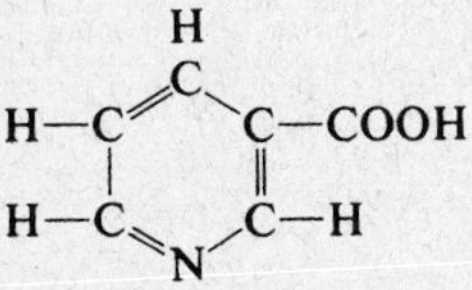

Nicotinic acid 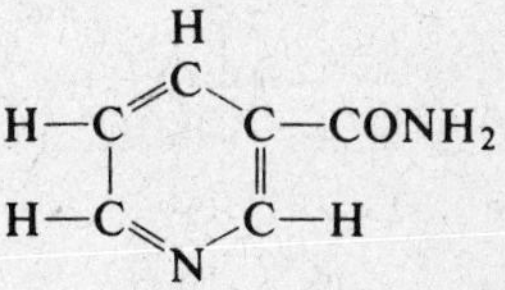                    Nicotinamide

Nicotinic acid in the body is converted to nicotinamide (niacinamide), which in turn is converted to nicotinamide adenine dinucleotide (NAD, DPN, or coenzyme I) and nicotinamide adenine dinucleotide phosphate (NADP, TPN, or coenzyme II).

Both of these coenzymes take part in many of the oxidation-reduction reactions in the cells.

A deficiency of this vitamin results in pellagra in man and blacktongue in dogs. Pellagra is sometimes called the "three D" disease because it is characterized by dermatitis, diarrhea, and dementia (mental disorders). It was formerly prevalent in the South, where many people subsisted largely on cornmeal, fat pork, and molasses. The disease is easily cured by giving foods rich in niacin, such as yeast, liver, lean meat, and wheat germ.

**Pyridoxine (Vitamin B$_6$).** Vitamin B$_6$ is a general name for three closely related compounds, pyridoxine, pyridoxal, and pyridoxamine. Like nicotinic acid and nicotinamide, they are derivatives of pyridine. Their structures are shown below.

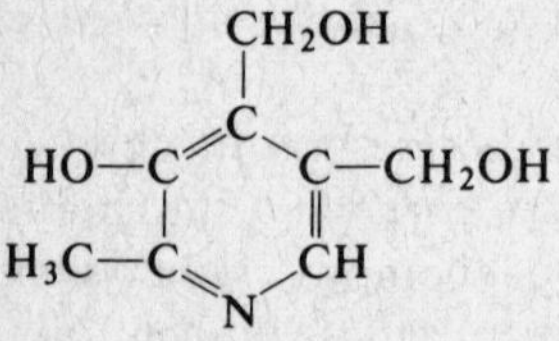

Pyridoxine 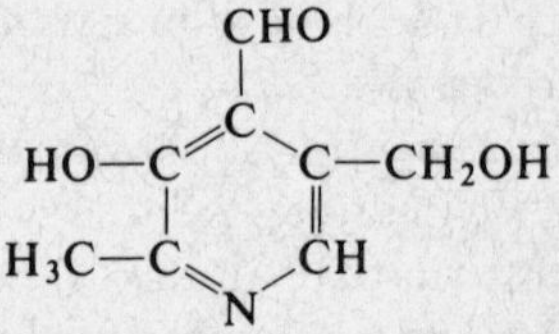                    Pyridoxal

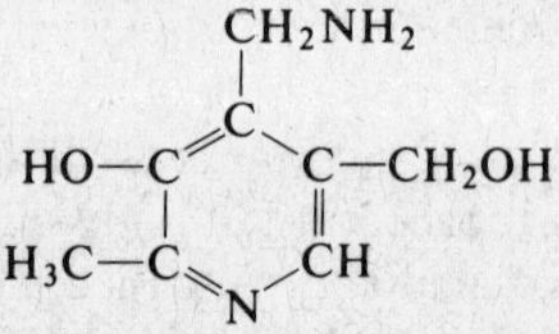

Pyridoxamine

Vitamin B$_6$ is essential for the normal nutrition of rats, pigs, dogs, and chickens, as well as for the normal growth of many microorganisms. A deficiency of this vitamin in rats produces redness and swelling of the paws, ears, nose, lips, chin, and side of the face. Such dermititis from vitamin B$_6$ deficiency has not been observed in man.

In the body, pyridoxine is converted to pyridoxal and pyridoxamine. Pyridoxal phosphate is a constituent of the coenzymes concerned with the transamination and decarboxylation of amino acids. Pyridoxine has been known to relieve certain symptoms of pellagra.

Egg yolk, yeast, wheat germ, liver, and fish are good sources of this vitamin.

**Pantothenic Acid.** Pantothenic acid is an aliphatic, long-chain hydroxy acid, consisting of substituted butane linked to an amino acid, beta alanine, as shown in the formula below.

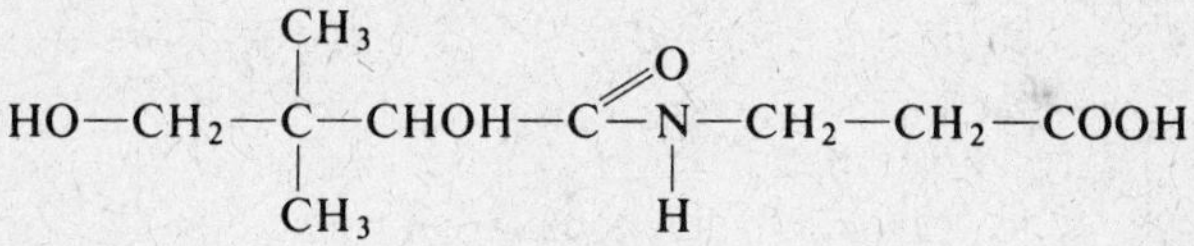

$$HO-CH_2-\overset{\overset{\displaystyle CH_3}{|}}{\underset{\underset{\displaystyle CH_3}{|}}{C}}-CHOH-\overset{\overset{\displaystyle O}{\parallel}}{C}-\overset{\underset{\displaystyle H}{|}}{N}-CH_2-CH_2-COOH$$

Pantothenic acid

Pantothenic acid is a constituent of coenzyme A and therefore is involved in the metabolism of pyruvate and fatty acids. It is essential to the normal growth of yeasts, bacteria, chicks, rats, dogs, and other animals. A deficiency of this vitamin in the diet of chicks results in dermatitis. In fur-bearing animals, such as rats, mice, and foxes, a lack of this vitamin causes graying of the hair. The natural color of the hair can be restored by giving the vitamin. There is no proof, however, that pantothenic acid can prevent the graying of the hair in humans.

Because pantothenic acid is widespread in nature, it was named from the Greek *pantos*, meaning everywhere. Yeast, liver, kidney, milk, and eggs are good sources of this vitamin.

**Biotin.** Biotin contains an imidazole ring fused to a thiophene ring as shown below.

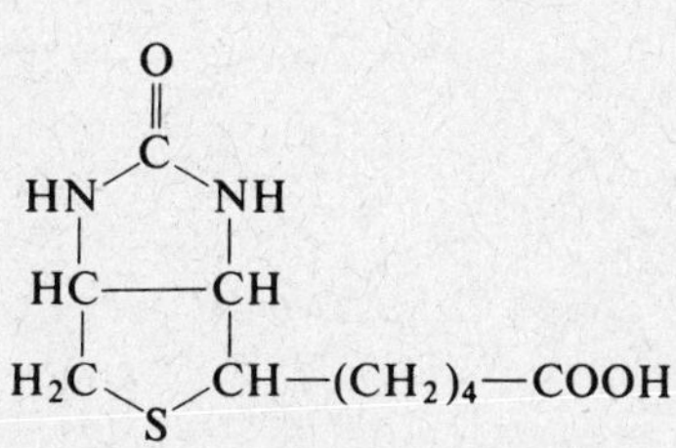

$$\begin{array}{c} \overset{\displaystyle O}{\underset{\displaystyle \parallel}{C}} \\ HN \qquad NH \\ HC\text{———}CH \\ H_2C \diagdown \qquad \diagup CH-(CH_2)_4-COOH \\ S \end{array}$$

Biotin

Biotin acts as a coenzyme in reactions which involve the transfer and splitting off of carbon dioxide, as in the synthesis of fatty acids and purines. Biotin is essential in the diet of animals and man. It is also required for the normal growth of yeast, bacteria, and fungi, and it is synthesized by bacteria in the intestines.

When raw egg white is fed to an animal, its protein *avidin* combines with biotin in the intestine to form an inactive complex which is not absorbed but is excreted in the feces. The ingestion of large amounts of raw egg white therefore can cause a biotin deficiency known as egg white injury, characterized by loss of hair and dermatitis. Cooked egg white does not produce egg white injury, since avidin is denatured by heat and therefore does not combine with biotin.

Among the symptoms of biotin deficiency in humans are skin rash, pallor, slow growth rate, lack of appetite, nausea, muscle pains, depression, and lassitude. These symptoms disappear when biotin is administered.

Peanuts, liver, kidney, and eggs are excellent sources of this vitamin.

**Inositol.** Inositol is a simple, nonnitrogenous cyclic compound called hexahydroxycyclohexane and has the following structure:

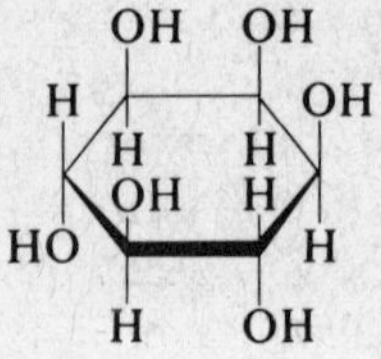

Inositol

Inositol is a constituent of the phosphoinositides and may act as a lipotropic agent in preventing fatty liver. It is also thought to regulate gastrointestinal motility. Inositol is sometimes called the mouse antialopecia factor because it cures the loss of hair produced by a deficiency of this vitamin. It has been found essential to the growth of mice.

Sources of inositol are muscle, liver, kidneys, and brain.

**Choline.** Choline is trimethylethanol ammonium hydroxide as shown by the following structure:

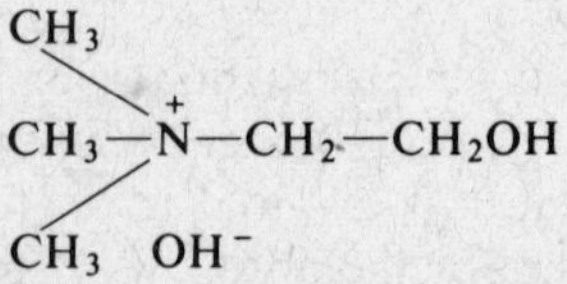

Choline

Choline is a constituent of lecithin, which is necessary for the transportation of fat in the body. The acetic ester of choline, acetylcholine, is liberated at the parasympathetic nerve endings and is involved in muscular contraction.

Choline deficiency produces fatty liver in dogs, rabbits, and rats; slipped tendons in chicks; and kidney hemorrhage in young rats immediately after weaning.

Sources of choline are egg yolk, brain, liver, and kidney.

**p-Aminobenzoic Acid.**   p-Aminobenzoic acid has the following simple structure, which is similar to that of sulfanilamide.

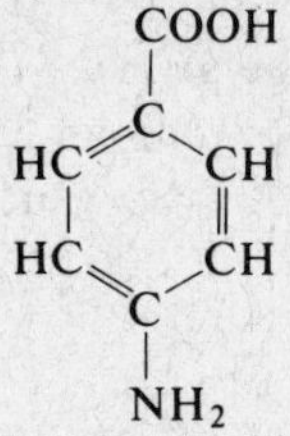

*p*-Aminobenzoic acid

Since *p*-aminobenzoic acid is essential for the growth of bacteria, and sulfanilamide is not, the latter has been used to check their growth. It appears that the administered sulfanilamide "competes" with *p*-aminobenzoic acid for a place on the active site of the bacterial enzyme system.   *p*-Aminobenzoic acid is also used by microorganisms for the synthesis of folic acid, though it cannot substitute for folic acid.

A deficiency of *p*-aminobenzoic acid in the diet retards the growth of young chicks and produces gray hair in black rats and mice, hence its other name, anti-grayhair factor.

**Folic Acid.**   Folic acid is a complex molecule consisting of pteridine, *p*-aminobenzoic acid, and glutamic acid.   For this reason it is sometimes called pteroylglutamic acid (PGA).

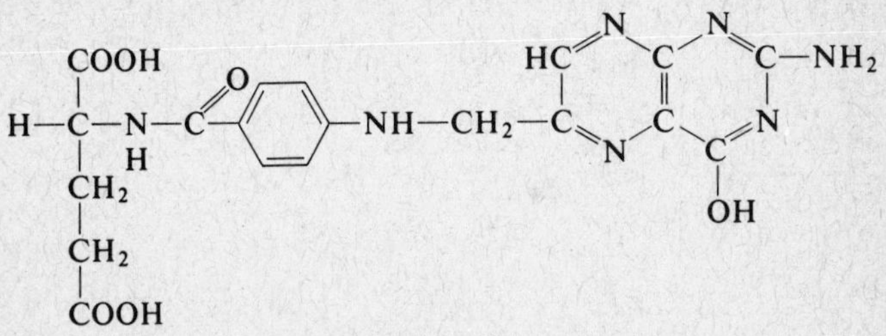

Folic acid

Folic acid serves as a carrier for one-carbon fragments in many methylation reactions in the body. It is essential for the growth of young animals and certain bacteria, for the normal maturation of erythrocytes in the bone marrow, and for the synthesis of purines, pyrimidines, and nucleoproteins. It is used in the treatment of macrocytic anemia, which is characterized by the presence of giant red corpuscles in the blood. It is also effective in treating sprue and leukopenia, a decreased number of white cells.

Folic acid was so-named because of its occurrence in the foliage of plants. However, its rich sources are liver, kidney, eggs, yeast, and soybeans.

**Vitamin $B_{12}$ (Cyanocobalamin).** Cyanocobalamin is a red crystalline compound containing an atom of cobalt linked to the four nitrogen atoms of a tetrapyrrole, to a cyanide group, and to a nucleotide. Its complex structure is shown below.

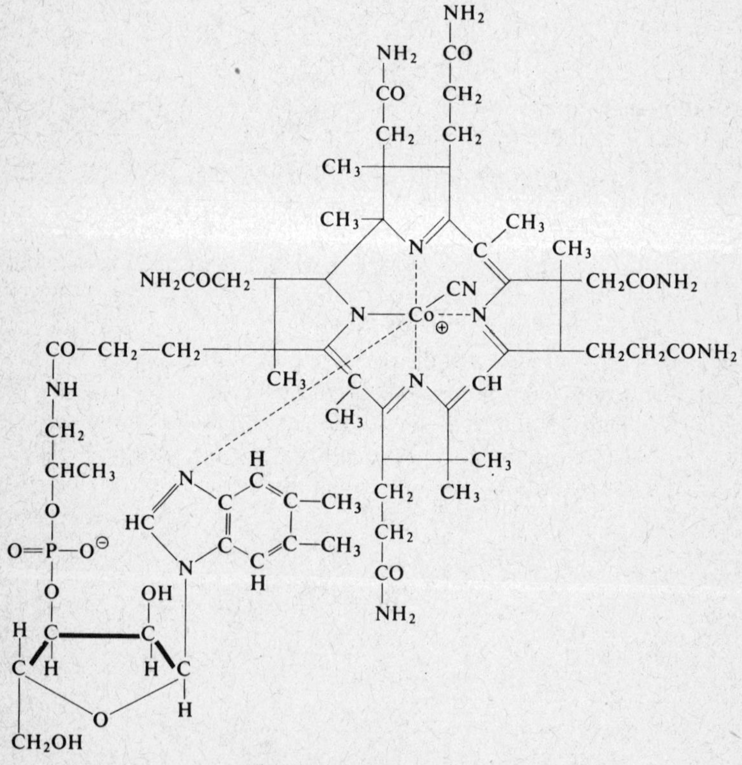

Vitamin $B_{12}$ (cyanocobalamin)

Cyanocobalamin is necessary for the biosynthesis of nucleic acids and certain amino acids. Like folic acid, this vitamin is involved in single-carbon reactions. Since vitamin $B_{12}$ stimulates red blood cell formation, it is used in the treatment of pernicious anemia. The vitamin acts as the *extrinsic factor* in the red cell maturation system. It cannot be absorbed in the intestine in the absence of the *intrinsic factor* of the gastric juice. Persons suffering from pernicious anemia generally do not produce sufficient intrinsic factor to enable the vitamin to be absorbed. To eliminate the need for the intrinsic factor, vitamin $B_{12}$ is given by injection.

The best source of vitamin $B_{12}$ is liver. Other sources include yeasts, meat, fish, milk, and eggs. The daily requirement is small, perhaps not more than 1 $\mu$g.

**Vitamin C (Ascorbic Acid).** The structure of ascorbic acid is closely related to that of glucose.

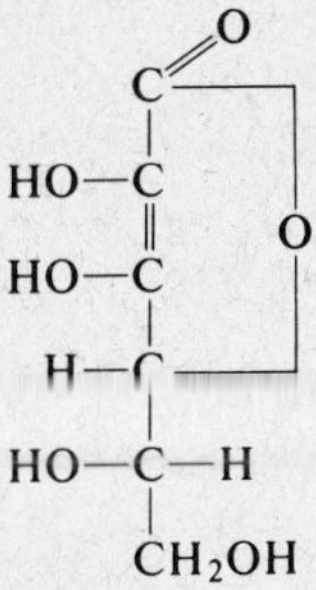

Ascorbic acid

One of the chief functions of vitamin C is the formation and maintenance of intercellular substances that make up cartilage, dentine of teeth, matrices of bone, and collagen of connective tissue.

Vitamin C is an active reducing agent and is therefore easily oxidized, especially in alkaline medium, at high temperatures, and in the presence of metals such as copper. When it is oxidized, it is converted to a ketone called dehydroascorbic acid. Since this reaction is reversible, it is suggested that vitamin C serves as a component of enzymes involved in oxidation-reduction reactions in the body tissues.

A deficiency of vitamin C in the diet produces scurvy. This disease is known to occur only in man, monkeys, and guinea pigs;

TABLE 29.1. SUMMARY

| Vitamin | Principal Synonyms | Structurally Related Chemical Derivatives |
|---|---|---|
| Vitamin A | Antixerophthalmic factor<br>Anti-infective factor<br>Vitamin $A_1$<br>Vitamin $A_2$ | $\alpha$-, $\beta$-, and $\gamma$-Carotene<br>Cryptoxanthin |
| Thiamine | Vitamin $B_1$<br>Antineuritic factor<br>Antiberiberi factor | Thiamine pyrophosphate<br>(cocarboxylase) |
| Riboflavin | Vitamin $B_2$<br>Vitamin G<br>Anticheilosis factor | Flavin mononucleotide (FMN)<br>Flavin adenine dinucleotide (FAD) |
| Nicotinic acid | Niacin<br>P-P factor<br>Antipellagra factor<br>Anti-blacktongue factor | Niacinamide (nicotinamide)<br>Coenzyme I (DPN or NAD)<br>Coenzyme II (TPN or NADP) |
| Pyridoxine | Vitamin $B_6$<br>Antidermatitis factor | Pyridoxal<br>Pyridoxamine |
| Pantothenic acid | Chick antidermatitis factor | Coenzyme A |
| Biotin | Vitamin H<br>Anti-egg white injury factor | |
| Inositol | Mouse antialopecia factor<br>Rat antispectacled eye factor | Phytin<br>Soybean cephalin |
| Choline | Growth factor | |
| $p$-Amino-benzoic acid | Anti-gray hair factor | |
| Folic acid | Pteroylglutamic acid<br>Anti-anemia factor | |
| Vitamin $B_{12}$ | Cyanocobalamin | Cobalamines |
| Vitamin C | Ascorbic acid<br>Antiscorbutic vitamin<br>Cevitamic acid | Dehydroascorbic acid |
| Vitamin D | Antirachitic vitamin<br>Vitamin $D_2$ or calciferol (irradiated ergosterol)<br>Vitamin $D_3$ (irradiated 7-dehydrocholesterol) | |
| Vitamin E | $\alpha$-, $\beta$-, and $\gamma$-Tocopherol<br>Antisterility factor<br>Fertility vitamin | |
| Vitamin K | Vitamin $K_1$<br>Antihemorrhagic vitamin<br>Coagulation vitamin | Vitamin $K_2$<br>Menadione |

| Main Biological Functions | Deficiency Symptoms | Main Sources | Av. Daily Requirements |
|---|---|---|---|
| Maintenance of healthy epithelial cells, formation of visual purple in retina | Xerophthalmia, night blindness, epithelial keratinization, failure of growth | Fish liver oil, butter, eggs, green and yellow vegetables | 5,000 I.U. 1.2 mg. vit. A acetate |
| Metabolism of carbohydrates | Beriberi, polyneuritis | Yeast, whole cereal grains, lean pork | 1–2 mg. |
| Oxidation in the tissues | Impaired growth, sore lips, cheilosis, vision defects | Yeast, liver | 2–3 mg. |
| Carbohydrate, fat, and protein metabolism | Pellagra, blacktongue disease in dogs | Yeast, wheat germ, liver | 10–30 mg. |
| Metabolism of amino acids | Dermatitis in rats | Yeast, rice polishings, eggs | 1.5 mg. |
| Transfer of acyl radical, metabolism | Dermatitis and gray hair in animals | Yeast, liver, peanuts, eggs | 2 mg. |
| Carboxylation and decarboxylation reactions | Alopecia in rats | Peanuts, liver, kidney, eggs | 0.2–0.6 mg. |
| Normal growth of mice, rats, and chicks | Alopecia in mice | Muscle, liver, kidney, brain | |
| Lipotropic function, synthesis of acetylcholine | Fatty liver in animals | Egg yolk, soybeans, brain, liver, kidney | |
| Maintenance of normal fur coat | Gray hair in rats and mice | Yeast, liver | |
| Red blood cell formation | Macrocytic anemia, growth retardation | Liver, kidney, dried beans, nuts, yeast | 0.5–1.0 mg. |
| Red cell formation, nucleic acid synthesis | Pernicious anemia | Liver, yeast | $\mu$g. quantities |
| Maintenance of normal intercellular tissue, oxidation-reduction reactions | Scurvy, hemorrhages of skin and gums, teeth defects | Citrus fruits, raw leafy vegetables, tomatoes, peppers | 70–80 mg. |
| Calcium and phosphorus metabolism, bone and tooth formation | Rickets, defective tooth structure | Fish liver oil, egg yolks, irradiated ergosterol, action of sunlight | 400–800 I.U. 10–20 $\mu$g. calciferol |
| Normal muscle metabolism and fertility in animals | Sterility, muscular dystrophy in animals | Wheat germ oil, corn and cottonseed oils, green leafy vegetables, egg yolk, meat | 6 mg. |
| Production of prothrombin | Hemorrhage, increased clotting time for blood | Alfalfa, kale, spinach, cabbage | 1–10 mg. |

other animals, such as rats, have the ability to synthesize vitamin C. The chief symptoms of scurvy in man are bleeding gums, loosening of the teeth, decalcification of the bones, hemorrhage under the skin, pains in the joints, and loss of weight. These symptoms disappear when the patient is given foods rich in vitamin C, such as citrus fruits, tomatoes, cabbage, green peppers, and raw leafy vegetables.

# 30. HORMONES

Hormones are organic compounds produced by the endocrine glands and carried by the blood to other glands, organs, and tissues where they exert their action. For this reason they have been called "chemical messengers." The word "hormone" comes from the Greek word meaning "to excite."

Like enzymes and vitamins, hormones are involved in the regulation of body activity. Enzymes directly catalyze reactions in the cells and tissues and vitamins serve as constituents of coenzymes; hormones are believed to exert their controlling action by influencing the activity of enzyme systems. Hormones can stimulate other glands to secrete their hormones. The necessary hormone balance of the body is dependent on feedback mechanisms.

Whereas enzymes are produced by glands that empty their secretions into ducts, hormones diffuse directly into the bloodstream. Thus the endocrine glands are known as the ductless glands or the glands of internal secretion. The endocrine glands are the pituitary, thyroid, parathyroids, pancreas, adrenals, ovaries, testes, and regions of the gastrointestinal tract. The location of these glands is shown in Fig. 30.1. The hormones produced by the endocrine glands are listed in the classification in Table 30.1.

### TABLE 30.1. CLASSIFICATION OF THE HORMONES

HORMONES OF THE PITUITARY
Anterior pituitary:
  *Growth or somatotropic hormone (GH)*
  Gonadotropic hormones
    *Luteinizing hormone (LH) or interstitial cell-stimulating hormone (ICSH)*
    *Follicle-stimulating hormone (FSH)*
  *Lactogenic hormone (prolactin)*
  *Thyrotropic or thyroid-stimulating hormone (TSH)*
  *Adrenocorticotropic hormone (ACTH)*
  *Diabetogenic hormone*
Posterior pituitary:
  *Vasopressin (pitressin)*
  *Oxytocin (pitocin)*

Table 30.1 (*continued*)

Intermediate pituitary (pars intermedia):
 *Melanocyte-stimulating hormone (intermedin)*
HORMONES OF THE THYROID
 *Thyroxine*
 *Triiodothyronine*
HORMONE OF THE PARATHYROID
 *Parathormone*
HORMONES OF THE PANCREAS
 *Insulin*
 *Glucagon*
HORMONES OF THE ADRENALS
 Adrenal medulla:
  *Epinephrine (adrenalin)*
  *Norepinephrine*
 Adrenal cortex:
  *Corticosterone*
  *Cortisone*
  *Hydrocortisone (cortisol)*
  *Aldosterone*
SEX HORMONES
 Male sex hormones (androgens)
  *Testosterone*
  *Androsterone*
 Female sex hormones (estrogens)
  Of the ovarian follicle:
   *Estrone*
   *Estradiol*
   *Estriol*
  Of the corpus luteum
   *Progesterone*
GASTROINTESTINAL HORMONES
 *Gastrin*
 *Secretin*
 *Cholecystokinin*
 *Enterogastrone*
 *Pancreozymin*

## HORMONES OF THE PITUITARY

The pituitary gland (hypophysis) is a small oval body situated at the base of the brain. It consists of three lobes: an anterior lobe, a posterior lobe, and a middle lobe called pars intermedia. The pituitary gland is sometimes referred to as the master gland of the body because its secretions have a controlling influence upon other endocrine glands.

### Anterior Lobe Hormones

Six hormones have been isolated in a highly purified form from the anterior lobe of the pituitary. These are: growth hormone,

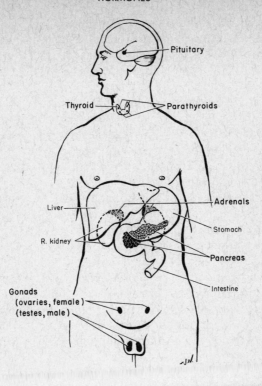

FIG. 30.1. Location of the endocrine glands. After Brooks: *Basic Facts of General Chemistry*. Philadelphia, W. B. Saunders Co., 1956.

luteinizing hormone, follicle-stimulating hormone, lactogenic hormone, thyrotropic hormone, and adrenocorticotropic hormone.

**Growth or Somatotropic Hormone (GH).** This hormone is a protein with a molecular weight of about 27,000. Its principal function is to stimulate skeletal growth and the formation of soft tissues. Underactivity of this hormone produces dwarfs (*dwarfism*), while overactivity produces giants (*giantism*) in children and *acromegaly* in adults, a condition characterized by progressive enlargement of the bones in the face, hands, and feet.

**Gonadotropic Hormones.** These hormones, also protein in nature, exert a stimulating effect on the gonads, ovaries in women and testes in men. The *luteinizing hormone* (LH), also known as the *interstitial cell-stimulating hormone* (ICSH), stimulates the formation of the corpus luteum and the production of progesterone in the female, and the production of testosterone in the male. The *follicle-stimulating hormone* (FSH) stimulates the growth of ovarian follicles in the female and the production of spermatozoa in the male.

**Lactogenic Hormone.** This hormone, also called *prolactin*, is a protein with a molecular weight of about 25,000. It initiates the secretion of milk by the mammary glands. During pregnancy, an inhibitor present in the placenta suppresses this hormone. When the placenta is separated after childbirth, prolactin is free to initiate and maintain lactation.

**Thyrotropic Hormone.** This hormone (TSH) is a protein with a molecular weight of about 10,000. It is required for normal thyroid development and the secretion of thyroxine. The production of this hormone drops when there is a high level of thyroid hormone in the blood.

**Adrenocorticotropic Hormone (ACTH).** This hormone is a polypeptide that contains 23 amino acids and has a molecular weight of 3,200. It regulates the proper functioning of the adrenal cortex and the production of cortical hormones. It is used in the treatment of rheumatoid arthritis and other diseases benefited by cortisone.

**Diabetogenic Hormone.** Injections of anterior pituitary extract into animals inhibit insulin and raise blood pressure. However, because the same effect is produced by the growth hormone and ACTH, the existence of a separate diabetogenic hormone is questionable.

### Posterior Lobe Hormones

The posterior lobe of the pituitary gland produces two known hormones, vasopressin and oxytocin, which are small polypeptides.

**Vasopressin (Pitressin).** Vasopressin is also called the *antidiuretic hormone* because it controls the formation of urine by stimulating the kidney tubules to reabsorb water. It is used to relieve excessive output of urine in diabetes insipidus. Vasopressin also raises the blood pressure by constricting the blood vessels and has been used to combat the low blood pressure of shock following severe injury or extensive surgery.

**Oxytocin (Pitocin).** This hormone stimulates the contraction of the smooth muscle of the uterus and has therefore been used in obstetrics to initiate labor, or after delivery to prevent post-partum hemorrhage.

### Pars Intermedia

This gland secretes the *melanocyte-stimulating hormone* (MSH), which consists of two straight-chain polypeptides, $\alpha$-MSH and $\beta$-MSH. These substances cause the migration of pigment granules in the skin of cold-blooded vertebrates from the center of the cell to the periphery, causing the skin to darken.

## HORMONES OF THE THYROID

The thyroid gland, located on either side of the trachea, consists of two lobes connected by a strip of tissue. It is filled with a colloidal material called *thyroglobulin* which is the source of the thyroid hormones.

**Thyroxine and Triiodothyronine.** The two most active thyroid hormones are thyroxine and triiodothyronine. The structural formulas below show that they are iodine derivatives of the amino acid tyrosine.

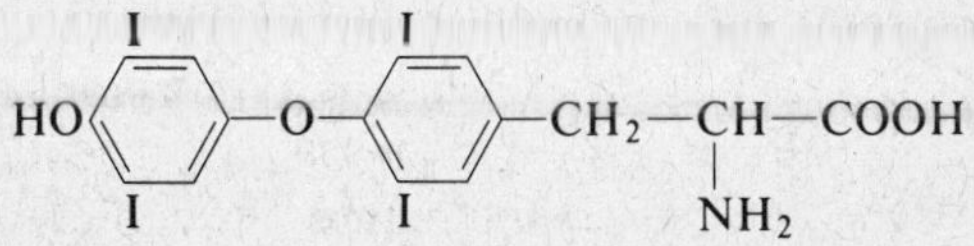

Thyroxine

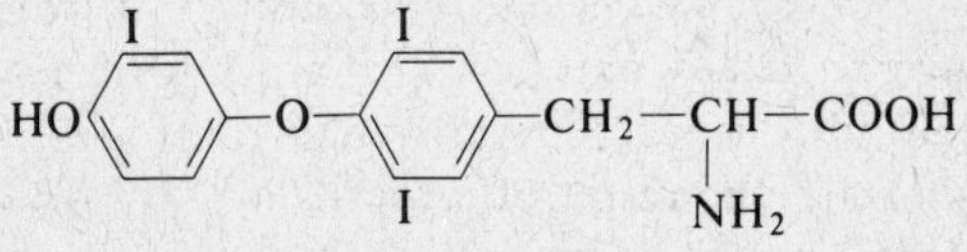

Triiodothyronine

The principal function of the thyroid hormones is to regulate the rate of metabolism in all cells of the body by controlling the oxidative reactions.

HYPOTHYROIDISM. In infants, abnormal development or underactivity of the thyroid gland produces cretins (*cretinism*), which are dwarfs of low mentality. If treatment with thyroid preparations is begun early, the child may develop normally.

Thyroid underactivity in adults may result in *myxedema*, which, like cretinism, is characterized by low pulse rate, low body temperature, thick dry skin, coarse hair, and physical and mental sluggishness. Myxedema also responds to treatment with thyroid gland preparations or thyroxine.

HYPERTHYROIDISM.   The symptoms of overactivity of the thyroid gland include increased metabolic rate, heart beat, and body temperature; exophthalmia or bulging of the eyes; nervousness and irritability; and loss of body weight.   There is also enlargement of the thyroid (*exophthalmic goiter*), which may be benefited by the surgical removal of part of the gland, the destruction of part of the thyroid tissue by radioactive iodine, or the use of antithyroid drugs to inhibit the synthesis of the thyroid hormones.

SIMPLE GOITER.   Simple goiter (*colloid goiter*, *endemic goiter*) refers to the enlargement of the thyroid gland caused by a lack of iodine in the food or drinking water.   The gland enlarges to compensate for the lack by increasing its glandular tissue.   This type of goiter can be prevented by using iodized salt or foods rich in iodine, such as seafood and kelp.

## HORMONE OF THE PARATHYROID

The parathyroids are four small glands attached to the thyroid gland. They produce a hormone called parathormone.

**Parathormone.**   Parathormone is a polypeptide with a molecular weight of 9,500. Its principal function is to maintain a normal level of calcium in the blood.   *Hypoparathyroidism* occurs when the glands are accidentally removed during thyroidectomy.   In this condition, there is decreased concentration of calcium in the blood, resulting in tetany and death.   Parathyroid tetany is best treated by administration of parathyroid hormone.   *Hyperparathyroidism* is usually the result of a tumor on the gland.   The observed symptoms include increased level of calcium in the blood, increased excretion of calcium in the urine, decalcification of the bones, and finally coma.   These conditions may be overcome by the surgical removal of a portion of the enlarged gland.

## HORMONES OF THE PANCREAS

The pancreas is a large, elongated gland lying behind the stomach.   Besides the pancreatic juice, which contains digestive enzymes, the pancreas produces the hormones insulin and glucagon.

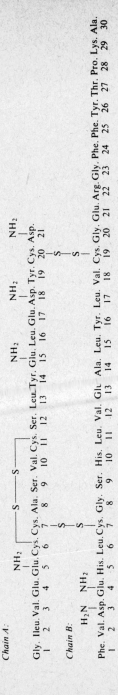

Chain A:

Gly. Ileu. Val. Glu. Glu. Cys. Cys. Ala. Ser. Val. Cys. Ser. Leu. Tyr. Glu. Leu. Glu. Asp. Tyr. Cys. Asp.
 1    2    3    4    5    6    7    8    9   10   11   12   13   14   15   16   17   18   19   20   21

Chain B:

Phe. Val. Asp. Glu. His. Leu. Cys. Gly. Ser. His. Leu. Val. Glu. Ala. Leu. Tyr. Leu. Val. Cys. Gly. Glu. Arg. Gly. Phe. Phe. Tyr. Thr. Pro. Lys. Ala.
 1    2    3    4    5    6    7    8    9   10   11   12   13   14   15   16   17   18   19   20   21   22   23   24   25   26   27   28   29   30

FIG. 30.2.    The structure of beef insulin.

**Insulin.** Insulin is produced in the beta cells of the islands (or islets) of Langerhans of the pancreas. It is a protein with a molecular weight of 12,000. It consists of two polypeptide chains joined together by cross-linkages of two disulfide bonds of the amino acid cystine. The structure of beef insulin is given in Fig. 30.2. Each amino acid is designated by the first three letters of its name.

The principal function of insulin is to control the oxidation of glucose in the body. Lack of insulin is the chief cause of diabetes mellitus, a disease characterized by increased blood sugar level, excretion of sugar in the urine, decreased storage of glycogen in the liver and muscle, and increased production of ketone bodies. Injection of insulin will produce a marked and rapid recovery from these conditions. Insulin cannot be taken by mouth because, as a protein, it would be hydrolyzed by the proteolytic enzymes of the digestive tract.

**Glucagon.** This hormone is produced in the alpha cells of the islands of Langerhans of the pancreas. It is a polypeptide with a molecular weight of 3,500. It raises the blood sugar level by increasing the rate of glycogen breakdown in the liver; thus its action is antagonistic to that of insulin.

## HORMONES OF THE ADRENALS

The adrenal or suprarenal gland is a small body attached to the top of each kidney. It consists of an inner portion called the *medulla* and an outer portion called the *cortex*.

### Hormones of the Medulla

The medulla secretes two hormones, epinephrine and norepinephrine, which are closely related in structure and action. The structural formula of epinephrine is given below.

$$HO-C_6H_3(OH)-CHOH-CH_2-N(H)-CH_3$$

Epinephrine

**Epinephrine (Adrenalin).** Epinephrine constricts arterioles, dilates bronchioles, stimulates heart muscle, relaxes smooth muscle, raises blood pressure, and increases blood sugar. It is the body's emergency hormone, for it mobilizes the energies of the body in times of emotional stress and tension, such as anger, fear, fight, or flight. It is used in surgery to check bleeding and to start heart action which has failed. It is also used as a decongestant in the treatment of asthma and hay fever.

**Norepinephrine.** This hormone differs from epinephrine in that the methyl group on the nitrogen atom is replaced by a hydrogen atom. It has functions similar to those of epinephrine, except that it does not affect carbohydrate metabolism or relax smooth muscle.

## Hormones of the Adrenal Cortex

The adrenal cortex secretes a group of hormones known as *cortin*, which are all steroids. The structures of the four most important corticosteroid hormones are given below.

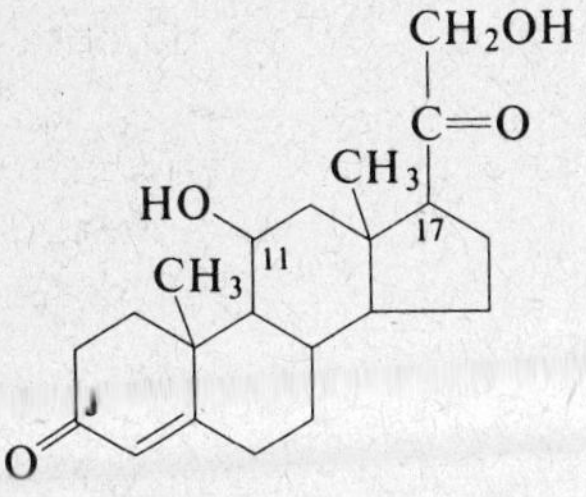

Corticosterone

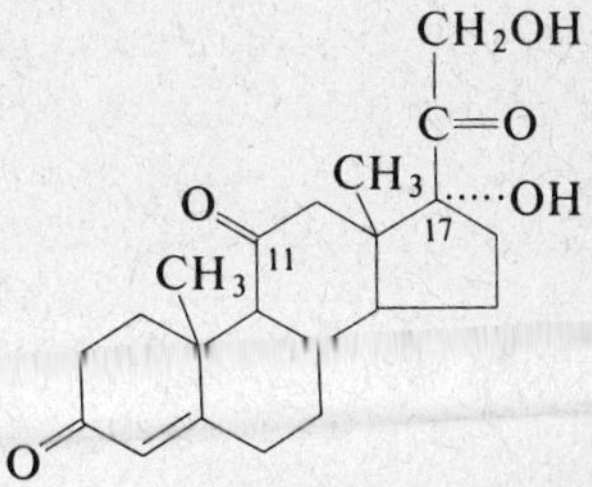

Cortisone

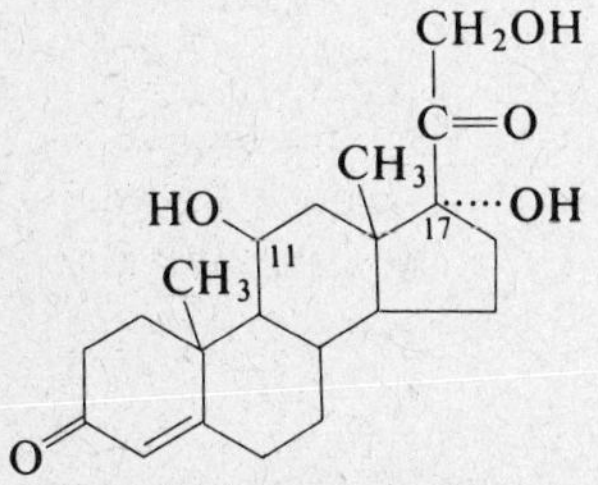

Hydrocortisone (cortisol)

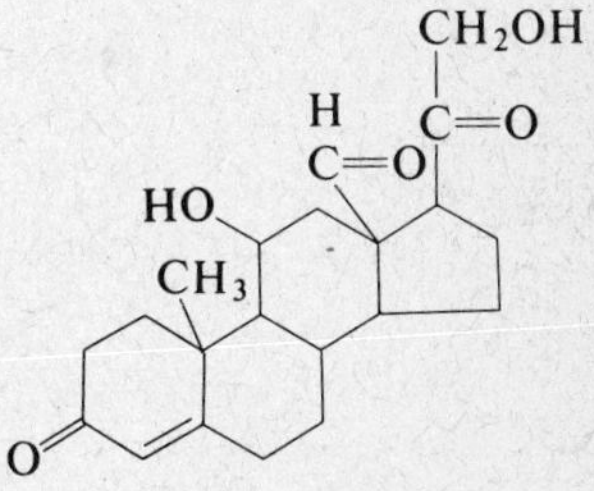

Aldosterone

The corticosteroid hormones may be classified into two groups. Those that are concerned with the metabolism of carbohydrates are called *glucocorticoids*, which include corticosterone, cortisone, and hydrocortisone or cortisol. The hormones that regulate sodium and potassium balance are called *mineralocorticoids*, the most important of which is aldosterone. The corticosteroid hormones are essential for life, since removal of the cortex from an animal usually results in death within ten days.

Hypofunction of the adrenal cortex, which may be caused by tuberculosis of the gland, results in *Addison's disease*. Its observed symptoms are loss of appetite, nausea, vomiting, bronze coloration of the skin, emaciation, and extreme weakness, the result of the low level of sugar and the high level of urea and other waste products in the blood. There is also an increased level of potassium and a decreased level of sodium in the blood. Sodium is excreted in the form of sodium chloride, and its excretion is accompanied by losses of water. The dehydration that results causes a decreased blood volume. The use of corticoid hormones will induce prompt remission of all these symptoms.

Cortisone and hydrocortisone have been used in the treatment of rheumatoid arthritis, rheumatic fever, asthma, and allergic diseases of the eye, skin, and mucosa. However, they do not cure these conditions and may have serious side effects.

Hyperfunction of the adrenal cortex, which may be caused by tumors, has a profound influence on sexual development. In children, it results in precocious growth and the early development of sexual characteristics. In women, it produces *adrenal virilism*, as manifested in the growth of a beard and the lowering of the pitch of the voice. At the same time, female characteristics are repressed.


## SEX HORMONES

The sex hormones of both the male and the female are all steroids. They bear a close chemical relationship to cholesterol, from which they are derived. These hormones are necessary for the proper development of the genital and accessory organs and of the secondary sex characteristics.

### Female Sex Hormones

The female sex hormones are formed in the ovaries. They include estrone, estradiol, estriol, and progesterone.

**Estrogens.** Estrone, estradiol, and estriol are called estrogens because they have the physiological property of inducing estrus,

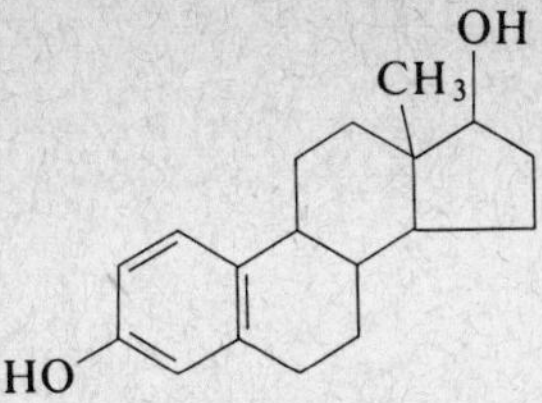

Estradiol

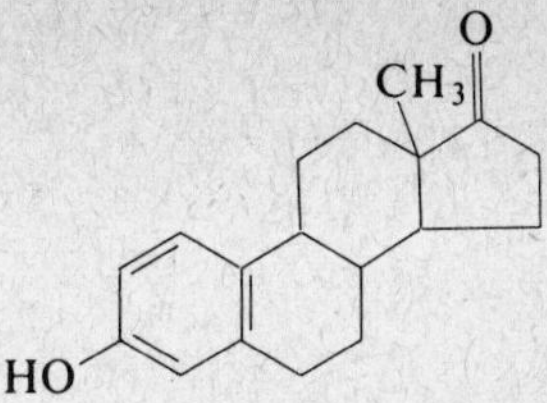

Estrone

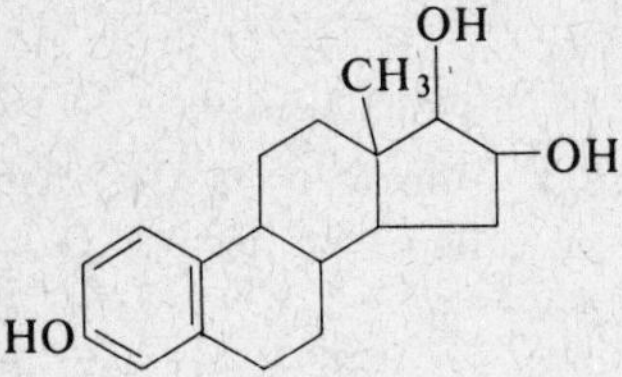

Estriol

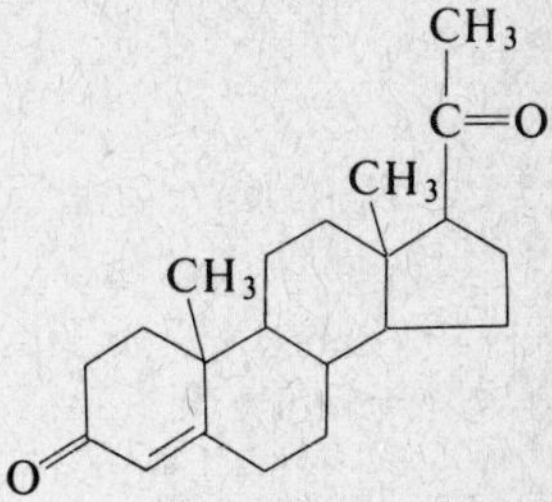

Progesterone

the urge for mating in lower animals. *Estradiol* is the principal estrogen; it is believed to give rise to estrone and estriol. Estradiol is produced by the follicle cells of the ovary under the influence of the follicle-stimulating hormone of the pituitary. The amount of estradiol rises from the fourth day of the menstrual cycle until about the fourteenth day when the matured ovum is expelled into the uterus. The principal function of the estrogen hormones is to promote the growth of the follicle and the development of ova and to bring about a proliferation and thickening of the endometrium (the lining of the uterus) for the reception of the fertilized ovum. They also help to develop the secondary female sex characteristics. Estrone and estradiol have been used for estrogen deficiency and for cancer of the prostate. A synthetic compound, *stilbestrol*, has the same effect as estradiol and may be administered orally.

**Progesterone.**  Progesterone is produced by the corpus luteum, the yellow body that develops in the follicle after the expulsion of the matured ovum. It prepares the endometrium of the uterus for the attachment and nourishment of the fertilized ovum and the development of the placenta. The amount of progesterone is maintained through pregnancy. If there is no fertilization of the ovum, the corpus luteum degenerates, production of progesterone ceases, and menstruation begins again about the twenty-eighth day.

## Male Sex Hormones

The male sex hormones are called *androgens*. They have structures similar to those of the estrogens.

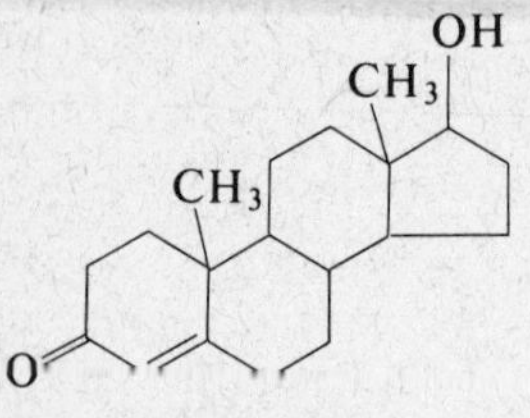

| Testosterone | Androsterone |
|---|---|

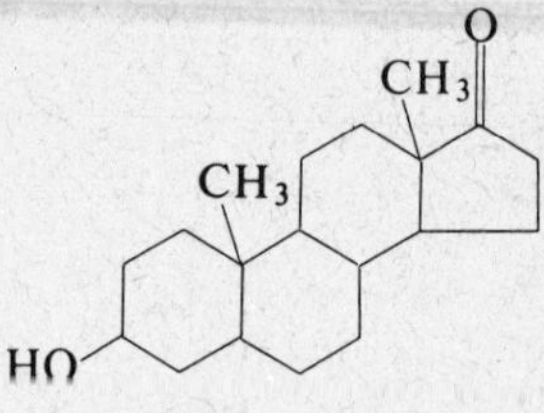

**Testosterone.** The most potent androgen is testosterone, which is produced in the interstitial cells of the testes. It is converted in the body into *androsterone*, which is excreted in the urine. The principal function of testosterone is to stimulate the development of the male reproductive organs and the male sex characteristics. Testosterone has been used for treating androgen deficiency (e.g., in eunuchs), breast cancer, and uterine disorders.

## HORMONES OF THE GASTROINTESTINAL TRACT

The five most important hormones of the gastrointestinal tract are gastrin, secretin, cholecystokinin, enterogastrone, and pancreozymin. They are either polypeptides or low molecular weight proteins.

**Gastrin.** This hormone is produced in the mucosa of the pyloric end of the stomach. The presence of food in the stomach causes it to diffuse into the bloodstream, which carries it to the cells of the stomach wall. There it stimulates secretion of gastric juice and causes contraction of the stomach muscles.

**Secretin.** Secretin is a polypeptide with a molecular weight of 5,000. It is secreted in the mucosa of the upper intestinal tract when acid chyme from the stomach enters the duodenum. The blood carries it to the pancreas, where it stimulates the flow of pancreatic juice.

**Cholecystokinin.** Cholecystokinin is also formed in the mucosa of the upper intestinal tract. It is released into the bloodstream when acid food or fat enters the duodenum. Its function is to stimulate the gallbladder to contract and empty its bile into the intestine.

**Enterogastrone.** This is another hormone that is produced in the intestinal mucosa. It is released into the bloodstream when a large amount of fat enters the small intestine. Its action is antagonistic to that of gastrin, for it reduces the flow of the gastric juice and the mobility of the stomach.

**Pancreozymin.** This hormone is also formed in the intestinal mucosa. It stimulates the production of pancreatic juice that is rich in enzymes.

# GLOSSARY

# GLOSSARY

**absolute temperature.**   A scale for measuring temperature that is obtained by adding algebraically 273 to degrees Centigrade.

**absolute zero.**   The temperature, 273 degrees below the zero of the Centigrade scale, at which all molecular motion ceases.

**absorption.**   The passage of digested foods from the intestine to the blood or lymph.

**acetone bodies.**   Substances formed in the liver by fat catabolism: acetone, acetoacetic acid, and beta-hydroxybutyric acid; also called ketone bodies.

**acetonuria.**   Acetone bodies in the urine; also called ketonuria.

**achlorohydria.**   A condition characterized by the absence of free hydrochloric acid in the gastric juice.

**acid.**   A compound which yields hydrogen ions in aqueous solution; a proton donor.

**acidosis.**   A condition in which the alkali reserve of the body is lowered by excessive acid production or imperfect acid excretion.

**acromegaly.**   A disease characterized by enlargement of the head, feet, and hands due to overactivity of the growth hormone in adulthood.

**activator.**   A metal ion that renders active an enzyme that is secreted in an inactive form.

**Addison's disease.**   A disease characterized by loss of strength, low blood pressure, and pigmentation of the skin, due to underactivity of the adrenal cortex.

**addition reaction.**   A reaction in which a reagent adds to a carbon-carbon double or triple bond.

**adrenal cortex.**   The outer portion of the adrenal gland.

**adrenal medulla.**   The inner portion of the adrenal gland.

**adsorption.**   A process in which molecules or ions adhere to the surface of a solid.

**albuminuria.**   Presence of albumin in the urine.

**alcohol.**   An organic compound containing the hydroxyl (OH) group.

**aldehyde.**   An organic compound containing the aldehyde (CHO) group.

**aldose.**   A sugar that is a polyhydric aldehyde.

**alicyclic compound.**   An organic compound that exhibits aliphatic characteristics but whose carbon atoms are in the form of a ring rather than an open chain.

**aliphatic compound.**   An organic compound of the open-chain structure.

**alkali.**   A water-soluble base that yields hydroxyl ions in aqueous solution.

**alkali reserve.**   The total amount of available buffering substances in the blood that neutralize excess acid.

**alkaline tide.**   A temporary rise in the alkalinity of the urine shortly after a meal, due to the increased secretion of acid gastric juice.

**alkaloids.**   Nitrogenous heterocyclic organic compounds of plant origin having powerful physiological effects.

**alkalosis.**   A condition in which the alkalinity of the blood and tissues is increased by excessive alkali intake or continued loss of acid.

**alkane.**   A saturated aliphatic hydrocarbon of the methane series.

**alkene.**   An unsaturated aliphatic hydrocarbon containing a double bond.

**alkyl group.**   A univalent radical derived from an alkane by removal of a hydrogen atom.

**alkyne.**   An unsaturated aliphatic hydrocarbon containing a triple bond.

**allotropism.**   The property of certain elements of existing in more than one form, as oxygen and ozone.

**alopecia.**   Loss of hair; baldness.

**alpha particle.**   The positively charged nucleus of the helium atom.

**amide.**   Any organic compound having the type formula $RCONH_2$.

**amine.**   An organic compound derived from ammonia by replacement of one or more of its hydrogen atoms by one or more hydrocarbon radicals.

**amino acid.**   An organic compound containing both the amino ($NH_2$) group and the carboxyl (COOH) group.

**amphoteric.**   Capable of acting both as an acid and as a base.

**anabolism.**   Constructive, or synthetic, metabolic processes taking place in the body.

**anaerobic.**   Living in the absence of free oxygen.

**analgesic.**   A remedy that relieves or removes pain.

**androgen.**   A male sex hormone.

**anhydremia.**   A deficiency of the fluid portion of the blood.

**anhydrous.**   Without water.

**anion.**   A negatively charged ion.

**anode.**   The positively charged electrode of an electrolytic cell.

**antimalarial.**   A substance that prevents malaria.

**antipyretic.**   A drug that lowers or prevents fever.

**antiseptic.**   A chemical substance that inhibits the growth of micro-organisms or destroys them.

**anuria.**   The lack of urine excretion.

**apoenzyme.**   The protein component of an enzyme.

**aromatic compounds.**   Cyclic organic compounds derived from benzene.

**atom.**   The smallest unit of an element that participates in chemical change.

**atomic number.** A number, characteristic of an element, that represents the number of protons in the nucleus of the atom. It indicates the location of an element in the periodic table.

**atomic weight.** The relative weight of an atom of an element referred to carbon, whose weight has been arbitrarily set at 12.01115.

**Avogadro's law.** The law which states that equal volumes of all gases, at the same temperature and pressure, contain the same number of molecules.

**Avogadro's number.** The number of molecules in one gram-molecular weight of any substance, $6.023 \times 10^{23}$.

**basal metabolic rate (B.M.R.).** The amount of heat produced by the body to maintain life when the body is in a state of physical, emotional, and digestive rest. It is expressed as kilocalories per hour per square meter of body surface.

**base.** A compound which yields hydroxide ions in aqueous solution; a proton acceptor.

**beriberi.** A disease caused by a lack of vitamin $B_1$ (thiamine) and marked by paralysis of the extremities and severe emaciation or swelling of the body.

**beta particle.** A negative electron given off by a radioactive substance.

**binary compound.** A compound composed of two elements per molecule.

**biochemistry,** or biological chemistry. The chemistry of living processes.

**boiling point.** The temperature at which the vapor pressure in a liquid equals the atmospheric pressure.

**Bowman's capsule.** The capsule in the kidney that envelops the glomerulus.

**Brownian movement.** The rapid oscillatory movement of small particles when suspended in water or other liquids.

**BTU.** British thermal unit, the quantity of heat required to raise the temperature of one pound of water one degree Fahrenheit. It is equal to 0.252 kilocalorie.

**buffer.** A substance that keeps the pH of a solution relatively constant in spite of the addition of considerable amounts of acid or base.

**calorie.** A small calorie (cal.) is the amount of heat required to raise the temperature of 1 gram of water 1°C. A large Calorie (Cal.) equals 1,000 small calories.

**carbohydrates.** Polyhydric aldehydes or polyhydric ketones, or substances which yield these upon hydrolysis. The principal carbohydrates are sugars, starches, cellulose, and glycogen.

**carbonyl group.** A bivalent organic radical (—CO—) occurring in aldehydes, ketones, acids, and their derivatives.

**carboxyl group.** A univalent organic radical ($-COOH$) which is the functional group of all the carboxylic acids.

**catabolism.** A breaking down process; destructive metabolism.

**catalyst.** A substance which speeds up, or less frequently retards, a chemical reaction, without being changed itself.

**cathode.** The negatively charged electrode of an electrolytic cell.

**cation.** A positively charged ion.

**cheilosis.** A disease characterized by reddened lips and fissures at the angles due to a deficiency of riboflavin.

**chemical change.** A change in which the composition of a substance is altered.

**chemistry.** The science that deals with the composition and properties of substances and their transformation from one form to another.

**chlorophyll.** The green coloring matter of plants, essential to the production of carbohydrates by photosynthesis.

**cholecystitis.** Inflammation of the gallbladder.

**chyme.** The semifluid material into which food is converted by gastric digestion.

**coagulation.** The process of changing a fluid into a thickened mass.

**coenzyme.** A nonprotein component of an enzyme that activates it.

**colloids.** Particles which are intermediate in size between crystalloids that form true solutions and suspensions that eventually settle.

**combining weight.** The weight of an element which will combine with 8 grams of oxygen, or 1.008 grams of hydrogen.

**combustion.** Rapid oxidation accompanied by evolution of heat and usually light.

**compound.** A substance composed of two or more elements united chemically in definite proportions by weight.

**corpus luteum.** A ductless gland developed within the ovary by the reorganization of a Graafian follicle following ovulation.

**covalence.** The combining of atoms by means of the sharing of electrons.

**crenation.** The shrinking of the red cells due to the passage of water from inside the cell to the surrounding hypertonic solution.

**cretinism.** A pathological condition due to hypothyroidism, characterized by dwarfism and idiocy.

**crystalloids.** Substances which, when dissolved in a liquid, will diffuse through a semipermeable membrane.

**Cushing's syndrome.** A condition caused by lesions of the pituitary gland in which there is adiposity of face, neck, and trunk; sexual dystrophy in females and impotence in males; dusky appearance of the skin.

**cyclic compound.** An organic compound that contains a closed chain or a ring of atoms.

**deamination.** A process involving the splitting off of the amino group from an amino acid.

**decarboxylation.** The process of removing carbon dioxide from a carboxylic acid.

**dehydration.** The removal of water from a substance.

**deliquescence.** The process whereby certain substances absorb sufficient water from the atmosphere to dissolve themselves.

**denaturation.** The process of altering the structure of a protein by physical or chemical means.

**density.** Mass per unit volume.

**deoxidation.** The process of removing oxygen from a compound.

**deuterium.** An isotope of hydrogen having twice the mass of ordinary hydrogen; also called heavy hydrogen.

**diabetes insipidus.** A disease in which there is an abnormal production of urine due to a deficiency of vasopressin.

**diabetes mellitus.** A disease in which the ability of the body to use sugar is impaired, and sugar appears abnormally in the urine.

**dialysis.** The separation of crystalloids from colloids in a solution by the diffusion of crystalloids through a semipermeable membrane.

**diffusion.** The process by which one substance distributes itself uniformly through another.

**digestion.** The process by which food is broken down into simple, soluble molecules small enough to pass through the intestinal wall.

**disaccharide.** A sugar that yields two molecules of simple sugar on hydrolysis.

**disinfectant.** A substance that destroys bacteria.

**dispersion.** Colloidal particles suspended in a liquid medium.

**distillation.** The evaporation of a liquid and the condensation of its vapor.

**diuretic.** A substance that increases the volume of urine.

**eclampsia.** Toxemia of pregnancy.

**edema.** A pathological state in which there is abnormal accumulation of fluid in the tissues or body cavities.

**efflorescence.** The spontaneous loss of water by a substance when exposed to the air.

**electrolysis.** The decomposition of a chemical compound by an electric current.

**electrolyte.** A substance which will conduct an electric current when in solution or when melted.

**electron.** An atomic particle carrying a unit charge of negative electricity, having a mass of 1/1837 of that of the proton.

**electrophoresis.** The migration of colloidal particles dispersed in a fluid, under the influence of an electric field.

**electrovalence.** The valence as determined by the electrons gained or lost by the elements reacting to form a compound.

**element.** One of the 103 fundamental forms of matter that cannot be separated into simpler substances by ordinary chemical means.

**emulsion.** A colloidal dispersion of a liquid in another liquid.

**endothermic reaction.** A chemical change in which heat is absorbed.

**energy.** The ability to do work.

**enzyme.** An organic catalyst formed in the living cell that influences the rate of chemical reactions in the body.

**equilibrium.** A state existing in a reversible reaction when the rates of the forward and reverse reactions are equal and the concentrations of the reactants and products are constant.

**erythrocyte.** A red corpuscle of the blood.

**ester.** An organic compound formed by the reaction between an acid and an alcohol.

**ether.** An organic oxide of the general formula R—O—R.

**exothermic reaction.** A chemical change in which heat is liberated.


**fat.** An ester formed from glycerol and fatty acids.

**fermentation.** The conversion of sugars into alcohol and carbon dioxide by the action of enzymes such as those in yeast and in bacteria.

**filtration.** The process of separating suspended particles from a liquid by means of a porous medium.

**fission.** The disintegration of the nucleus of a heavy atom with the liberation of a large amount of energy.

**formula.** An expression of the constituents of a compound by symbols.


**gamma ray.** Similar to an X ray, forming part of the radiation of a radioactive substance.

**gel.** A semirigid colloid.

**globulin.** A simple protein, insoluble in water but soluble in dilute salt solutions.

**glomerulus.** A compact cluster of capillaries in the kidney.

**glycogenesis.** The formation in the liver of glycogen from glucose.

**glycogenolysis.** The breaking down in the liver of glycogen into glucose.

**glycolysis.** The conversion of glucose to lactic acid by an anaerobic process.

**glycoprotein.** A protein containing a carbohydrate group.

**glycosuria.** A condition in which an abnormal amount of glucose is present in the urine; also referred to as glucosuria.

**goiter.** An enlargement of the thyroid gland.

**gonadotropic hormones.** Hormones formed in the pituitary or in the placenta which affect the activity of the ovary or testis.

**gonads.** Male or female sex glands in which the germ cells develop.

**gout.** Painful inflammation of the joints in the feet and hands, especially in the great toe.

**gram-atomic weight.** One atomic weight of an element expressed in grams.

**gram-molecular weight.** One molecular weight of a compound expressed in grams.

**Graves' disease.** A disease characterized by an enlarged thyroid, rapid pulse, and increased basal metabolism due to excessive thyroid secretion.


**half-life.** The length of time required for one half of a radioactive substance to disintegrate.

**halide.** A compound composed of two elements, one of which is a halogen.

**halogens.** A family of elements consisting of fluorine, chlorine, bromine, and iodine.

**hematuria.** The presence of blood in the urine.

**hemoglobin.** The respiratory pigment of the blood.

**hemoglobinuria.** The presence of hemoglobin in the urine.

**hemolysis.** The rupturing of the red blood cells with liberation of hemoglobin.

**heterocyclic compounds.** Cyclic compounds in which the ring system of the molecule contains other elements than carbon.

**hexose.** A simple sugar containing six carbon atoms in the molecule.

**holoenzyme.** The complete enzyme consisting of apoenzyme and co-enzyme.

**hormones.** Chemical substances secreted by endocrine glands and carried by the blood to other glands, organs, and tissues whose activities they regulate or coordinate.

**hydrate.** A compound containing water of crystallization.

**hydride.** A compound containing a negatively charged hydrogen, as sodium hydride, $NaH$.

**hydrocarbons.** Organic compounds containing hydrogen and carbon only.

**hydrogenation.** A chemical reaction in which hydrogen is added to a compound.

**hydrolysis.** Chemical decomposition by which a compound is resolved into other compounds by taking up the elements of water.

**hyperacidity.** An excess of acidity, as of the gastric juice.

**hyperchlorhydria.** A condition caused by excess hydrochloric acid in the gastric juice.

**hyperglycemia.** A condition caused by an increase of blood sugar above the normal value.

**hyperinsulinism.** A condition caused by excessive production of insulin by the pancreas.

**hyperparathyroidism.** A condition caused by overactivity of the parathyroid gland.

**hyperthyroidism.** A condition caused by overactivity of the thyroid gland.

**hypertonic solution.** A solution of higher osmotic pressure than another with which it is compared.

**hypoacidity.** Deficient or subnormal acidity, as of the gastric juice.

**hypochlorhydria.** A condition caused by a decrease in the secretion of hydrochloric acid in the gastric juice below the normal level.

**hypoglycemia.** A condition caused by a decrease of blood sugar below the normal value.

**hypoparathyroidism.** A condition caused by diminished activity of the parathyroid gland.

**hypothyroidism.** A condition caused by diminished activity of the thyroid gland.

**hypotonic solution.** A solution of lower osmotic pressure than another with which it is compared.

**immiscible.** Incapable of being mixed, as oil and water.

**indicator.** A compound which changes color with changes in the hydrogen ion concentration (pH) of a solution.

**iodine number.** The number of grams of iodine required to saturate 100 grams of fat.

**ion.** An electrically charged atom or radical.

**ionization.** The separation of an electrolyte into charged ions in solution.

**isoelectric point.** The pH at which a substance is electrically neutral or at its minimum ionization.

**isomers.** Compounds which have the same molecular formula but different structural formulas.

**isotonic solution.** A solution having the same osmotic pressure as another with which it is compared.

**isotopes.** Atoms of the same element having the same atomic number but different atomic weights.

**jaundice.** A disease condition due to the presence of bile pigments in the blood, characterized by yellowness of the skin.

**keratinization.** A disease of the skin characterized by abnormal amounts of keratin, a dead skin layer.

**ketone.** An organic compound containing the carbonyl group attached to two organic radicals.

**ketone bodies.** See Acetone bodies.

**ketonuria.** The presence of ketone bodies in the urine.

**ketose.** A simple sugar that is a polyhydric ketone.

**ketosis.** See Acidosis.

**kindling temperature.** The lowest temperature at which a substance bursts into flame.

**kinetic energy.** Energy in motion.

**lactosuria.** The presence of lactose in the urine.

**latent heat.** The heat absorbed in the changing of a substance from solid to liquid or from liquid to gas.

**leukocyte.** A white corpuscle of the blood.

**lipid.** A fat or fatlike substance having a greasy feeling, which is insoluble in water but soluble in organic solvents.

**lipoprotein.** A protein in combination with a lipid.

**lymph.** A clear yellowish alkaline fluid that contains white blood cells.

**macrocytic anemia.** An anemia characterized by a predominance of macrocytes (abnormally large red blood cells).

**matter.** Anything that occupies space and has mass.

**metabolism.** Chemical changes that take place in the tissues of the body.

**micron.** One-millionth of a meter.

**millimicron.** One-thousandth of a micron.

**miscible.** Capable of being mixed.

**mixture.** An aggregate of two or more substances which are not chemically combined and which exist in no fixed proportion to each other.

**molal solution.** A solution containing one mole of the solute in 1,000 grams of the solvent.

**molar solution.** A solution which contains 1 gram-molecular weight of solute in 1 liter of solution.

**mole.** The molecular weight of a substance expressed in grams.

**molecular weight.** The sum of the atomic weights of all the atoms in a molecule.

**molecule.** The smallest particle of a compound that can exist independently.

**monosaccharide.** A simple sugar that cannot be decomposed by hydrolysis, e.g., glucose and fructose.

**multiple myeloma.** A disease characterized by the development of tumors from bone-marrow cells and the presence of Bence-Jones protein in the urine.

**mydriatic.** A drug that dilates the pupil of the eye.

**myxedema.** A disease caused by underactivity of the thyroid gland.

**nascent.** The condition of an element that has just been released in the monatomic state in a chemical reaction.

**nephritis.** Inflammation of the kidneys.

**nephron.** The unit of the kidney that is involved in urine formation, consisting of glomerulus, Bowman's capsule, and tubules.

**nephrosis.** Degenerative or retrogressive renal lesions, as distinct from inflammation (nephritis).

**neutralization.** The reaction between an acid and a base to form a salt and water.

**neutron.** A neutral particle existing in the nucleus of an atom.

**noble gases.** A family of elements consisting of helium, neon, argon, krypton, xenon, and radon.

**nonelectrolyte.** A compound whose water solution does not conduct an electric current.

**normal solution.** A solution which contains 1 gram-equivalent weight of the solute in 1 liter of the solution.

**nucleic acid.** An organic compound composed of phosphoric acid, ribose or deoxyribose, and purine and pyrimidine bases.

**nucleoprotein.** A protein in combination with a nucleic acid.

**nucleoside.** A purine or pyrimidine base linked to ribose or deoxyribose.

**nucleotide.** A phosphoric ester of a nucleoside.

**nucleus.** The positively charged center of the atom containing protons and neutrons.

**olefins.** An unsaturated aliphatic hydrocarbon containing a double bond.

**oliguria.** A diminished excretion of urine.

**organic chemistry.** The chemistry of carbon compounds.

**osmosis.** The passage of fluid from a less concentrated solution to a more concentrated solution through a semipermeable membrane.

**osteitis fibrosa cystica.** A disease due to parathyroid hyperfunction with excessive urinary excretion of calcium and phosphorus, characterized by thinning of the bone, often in the form of cysts.

**osteomalacia.** A disease of adults, characterized by gradual softening of various bones, due to deficiency of calcium, phosphorus, and vitamin C.

**oxidation.** Combination of a substance with oxygen; the increase in valence toward the positive; the loss of electrons.

**oxide.** A compound of oxygen and some other element.

**oxidizing agent.** A substance that brings about oxidation of another substance, itself being reduced in the process.

**oxygenation.** Oxygen combining loosely with a substance, as in the formation of oxyhemoglobin.

**oxyhemoglobin.** The compound formed when hemoglobin unites with oxygen.

**parathyroids.** Small endocrine glands situated near the thyroid.

**pellagra.** A disease caused by a deficiency of niacin, or nicotinic acid, characterized by dermatitis, diarrhea, and dementia.

**pentose.** A simple sugar having five carbon atoms in the molecule.

**peptide linkage.** The $-\overset{\overset{\text{O}}{\|}}{\text{C}}-\overset{\overset{\text{H}}{|}}{\text{N}}-$ linkage which unites the amino acids in the protein molecule.

**periodic law.** The law that the properties of the elements are periodic functions of their atomic numbers.

**periodic table.** A table illustrating the periodic system in which the chemical elements arranged in the order of their atomic numbers are shown in related groups.

**pernicious anemia.** A disease produced by a decrease in the number of the red blood cells.

**pH.** A symbol denoting the logarithm of the reciprocal of hydrogen ion concentration.

**phosphatides.** See phospholipids.

**phospholipids.** Fatlike substances which can be broken down to fatty acids, glycerol, phosphoric acid, and sometimes also a nitrogenous base.

**photosynthesis.** The process by which the chlorophyll of green plants catalyzes the formation of carbohydrates from carbon dioxide and water through the action of sunlight.

**physical change.** A change in the condition or state of a substance; its composition is not altered.

**pituitary.** A small oval endocrine gland attached to the base of the brain whose hormones regulate other endocrine glands.

**plasma.** The liquid part of the blood, as distinguished from the corpuscles.

**platelet.** A minute body found in the blood, essential to the coagulation process.

**polycythemia.** A condition characterized by an increased number of erythrocytes.

**polyhydric.** Having a number of hydroxyl groups.

**polymerization.** The linking together of many like molecules to form a larger one, a polymer, which has the same percentage composition as the smaller one but different properties.

**polyneuritis.** Multiple inflammation of nerves.

**polypeptide, or peptide.** A compound composed of two or more amino acids.

**polysaccharide.** A carbohydrate containing more than three monosaccharide units per molecule, such as starch, glycogen, and cellulose.

**polyuria.** Excessive urine secretion.

**potential energy.** Energy that is due to position and not to motion; stored energy.

**precipitate.** An insoluble solid that separates from solution.

**proenzyme.** The inactive form of an enzyme.

**prosthetic group.** The nonprotein cofactor attached to an enzyme.

**protamine.** A basic, simple protein.

**protein.** A substance of high molecular weight composed of carbon, hydrogen, nitrogen, and sometimes sulfur or iodine, that yields upon hydrolysis amino acids or their derivatives.

**proteinuria.** The presence of protein in the urine.

**proton.** A subatomic particle carrying a unit of positive charge.

**purgative.** A substance that causes evacuation of the bowels.

**radical.** A group of atoms that behave as a unit in a chemical reaction.

**radioactivity.** The spontaneous disintegration of an atom, with the emission of alpha, beta, and gamma rays.

**radioisotope.** A radioactive isotope, often used as a tracer in scientific research.

**reduction.** The removal of oxygen from a compound; the gain of electrons by a substance or a decrease in its valence.

**rickets.** A disease of childhood characterized by softening of the bones due to a lack of vitamin D.

**salt.** A compound consisting of a positive ion other than hydrogen and a negative ion other than the hydroxyl ion; the product of the reaction of an acid with a base.

**saponification.** The reaction between a fat and a base to form soap and glycerol, or between an ester and a base to form a salt and an alcohol.

**saponification number.** The number of milligrams of potassium hydroxide required to saponify one gram of fat.

**saturated compound.** An organic compound in which all valences are satisfied and which does not contain double or triple bonds.

**saturated solution.** A solution that contains all the solute it can hold at a given temperature and pressure.

**scurvy.** A disease caused by a lack of vitamin C in the diet.

**sedative.** A drug that allays nervousness, irritability, or excitement.

**semipermeable membrane.** A membrane that allows water and crystalloids to pass through but holds back colloids.

**serum.** Plasma minus fibrinogen, or whole blood minus cells and fibrinogen.

**soap.** The metallic salt of a high fatty acid.

**solute.** A substance dissolved in a solvent.

**solution.** A homogeneous mixture of two or more substances.

**solvent.** A substance in which a solute is dissolved.

**specific gravity.** The ratio of the weight of a given volume of a substance to the weight of an equal volume of water.

**specific heat.** The quantity of heat in calories required to raise the temperature of one gram of a substance one degree Centigrade.

**spontaneous combustion.** The ignition of a substance as the result of the accumulated heat of slow oxidation.

**stimulant.** A substance that temporarily quickens some vital process or functional activity.

**structural formula.** A formula which shows the arrangement of the atoms in the molecule.

**substitution reaction.** A chemical reaction in which one or more elements or radicals in a compound are replaced by other elements or radicals.

**suspension.** A system consisting of small particles dispersed in a liquid. The particles will settle out slowly upon standing.

**symbol.** A one- or two-letter abbreviation of the name of an element.

**syneresis.** The shrinking of a gel, with the expulsion of water or other liquid from it.

**synthesis.** The construction of a compound by the union of elements or simpler compounds.

**thyroid.** The endocrine gland lying on either side of the windpipe or trachea in the neck.

**tincture.** A solution of a medicinal substance in alcohol.

**titration.** The process of determining the quantity of a substance in a solution by adding a measured volume of a standard solution until the desired reaction has been effected.

**tranquilizer.** A drug that has a calming effect on overactive or disturbed patients.

**transamination.** The exchange between the amino group of an amino acid and the keto group of a keto acid, resulting in the formation of a new amino acid and a new keto acid.

**transmutation.** The conversion of one element into another in a nuclear reaction.

**Tyndall effect.** The reflection of a beam of light by the dispersed particles of a colloidal solution, making visible the path of the light.

**unsaturated compounds.** Organic compounds containing double or triple bonds and capable of forming addition products.

**unsaturated solution.** A solution containing less solute than the amount needed to make a saturated solution.

**uremia.** A pathological condition resulting from the accumulation in the blood of urinary constituents, such as urea, as a result of dehydration or dysfunction of the kidney.

**valence.** A number that represents the combining power of an element or a radical.

**valence electrons.** The electrons located in the outermost orbit of an atom.

**villi.** The tiny finger-like projections on the mucus membrane of the small intestine, containing blood and lymph vessels for the absorption of digested foodstuffs

**vitamin.** A food factor that does not supply energy, but is essential for proper metabolism.

**water of crystallization.** Water present in the crystal of a hydrate.

**wax.** An organic ester of a high molecular weight fatty acid and a high molecular weight alcohol.

**xerophthalmia.** A disease characterized by abnormal dryness of the eyeball due to a lack of vitamin A in the diet.

**zwitterion.** A dipolar ion carrying both a positive and a negative charge.

**zymogen.** An inactive enzyme; a proenzyme.

# SELF-SCORING TEST

# SELF-SCORING TEST

This self-scoring test is designed with the following aims in mind:

1. To aid the student in evaluating the achievement of his study of the course.

2. To help the student discover his mistakes and make appropriate corrections accordingly. In this sense, the test serves as a review.

3. To give the student the "feel" of the test, so that when a similar test or a standardized test is given by the instructor, the student will face it with more confidence.

With the exception of a few filling-in and matching questions, the questions are of the multiple choice type and require only one answer.

The answers to the questions are given at the end of the test.

The questions in this test follow the order of the chapters. The specific questions belonging to each chapter are indicated in the following tabulation:

| CHAPTER | QUESTIONS | CHAPTER | QUESTIONS |
|---------|-----------|---------|-----------|
| 1 | 1–3 | 16 | 95 100 |
| 2 | 4 7 | 17 | 104–107 |
| 3 | 8 16 | 18 | 108–110 |
| 4 | 17–26 | 19 | 111–118 |
| 5 | 27–33 | 20 | 119,120 |
| 6 | 34–40 | 21 | 121,122 |
| 7 | 41–44 | 22 | 123–127 |
| 8 | 45–48 | 23 | 128–130 |
| 9 | 49–55 | 24 | 131,132 |
| 10 | 56–62 | 25 | 133–135 |
| 11 | 63–72 | 26 | 136–137 |
| 12 | 73–75 | 27 | 138,139 |
| 13 | 76–83 | 28 | 140,141 |
| 14 | 84–90 | 29 | 142–146 |
| 15 | 91–94 | 30 | 147–150 |

1. Classify the following as elements (E), compounds (C), or mixtures (M):
   _____ a. Water
   _____ b. Air

_____ c. Carbon

_____ d. Salt

_____ e. Milk

2. Classify the following as physical (P) or chemical (C) changes:

    _____ a. Melting of snow

    _____ b. Salt dissolved in water

    _____ c. Rusting of iron

    _____ d. Milk turned sour

    _____ e. Water turned to steam

3. What is 72°F. in degrees Centigrade?

    a. 8°C.

    b. 17.6°C.

    c. 22.2°C.

    d. 40°C.

    e. 57.6°C.

4. Check the correct statement:

    a. Electrons are present in the nucleus.

    b. The atomic number of an element is equal to the number of neutrons in the nucleus.

    c. The atomic number of an element is equal to the number of valence electrons.

    d. Neutrons have practically the same weight as the protons.

    e. The mass of an element is due principally to the electrons.

5. Check the correct statement:

    When two elements react, there is

    a. loss or gain of neutrons.

    b. loss or gain of protons.

    c. loss or gain or sharing of valence electrons.

    d. noticeable loss or gain in weight.

    e. no change in their energy content.

6. Which of the following has more protons in the nucleus than electrons outside the nucleus?

    a. A helium atom

    b. A magnesium atom

    c. A magnesium ion

    d. A sulfur atom

    e. A sulfide ion

7. Isotopes are atoms that have the

    a. same atomic weight but different atomic numbers.

    b. same atomic number but different atomic weights.

    c. same number of neutrons but a different number of protons.

    d. same number of protons but a different number of electrons.

    e. same atomic weight but different properties.

8. Check the incorrect statement:

a. The properties of the elements are periodic functions of their atomic numbers.

b. Elements belonging to the same group have the same number of valence electrons.

c. The nonmetals are located in the upper left corner of the table.

d. The elements in a given group become more metallic with increasing atomic number.

e. The rare earth elements resemble each other more closely than other groups of elements.

9. Which one of the following scientists had the most to do with the periodic table?

    a. Cavendish

    b. Madam Curie

    c. Lavoisier

    d. Mendeleev

    e. Priestley

Which element in the parentheses most closely resembles the first element?

10. _____ O (H, N, Cl, S)

11. _____ Cl (Na, Br, Al, H)

12. _____ K (Ca, Na, Fe, Hg)

13. _____ He (F, N, Ne, H)

14. _____ Mg (Ca, Al, Ag, Li)

15. The elements Li, Na, K, Rb, and Cs are known as the

    a. Transition metals

    b. Alkali metals

    c. Alkaline earths

    d. Heavy metals

    e. Noble metals

16. Write the letter of the most appropriate item from Column (2) in the blank at the left of the corresponding Column (1).

| (1) | (2) |
|---|---|
| _____ a. Halogens | A. Group O |
| _____ b. Noble gases | B. Group IA |
| _____ c. Transition elements | C. Group IIIB |
| _____ d. Light metals | D. Group VIIA |
| _____ e. Rare earths | E. Group VIII |

17. Alpha particles are electrically charged

    a. Hydrogen atoms

    b. Neutrons

    c. Helium atoms

    d. X rays

    e. Deuterons

18. A radioactive atom having an atomic number 90 and atomic weight 234 loses a beta particle. Which of the following structures will the resulting atom have?

|      | Atomic number | Atomic weight |
|------|---------------|---------------|
| a.   | 88            | 234           |
| b.   | 88            | 232           |
| c.   | 89            | 233           |
| d.   | 90            | 233           |
| e.   | 91            | 234           |

19 The particle that initiates fission reactions is
   a. Electron
   b. Proton
   c. Ion
   d. Neutron
   e. Positron

20. Radioactive iodine, $^{131}I$, has a half-life of 8 days. If we have 100 micrograms of this isotope at one instant, how many micrograms will remain at the end of 16 days?
   a. Zero
   b. 25
   c. 33.3
   d. 50
   e. 100

21. Which one is NOT an instrument for detecting radiation?
   a. Electroscope
   b. Geiger counter
   c. Film badge
   d. Cyclotron
   e. Wilson cloud chamber

Complete the following nuclear reactions:

22. $^{234}_{90}Th \rightarrow \, ^{234}_{91}Pa + ?$
23. $? + \, ^{1}_{0}n \rightarrow \, ^{239}_{92}U$
24. $^{14}_{7}N + ? \rightarrow \, ^{17}_{8}O + \, ^{1}_{1}H$
25. $^{32}_{16}S + \, ^{1}_{0}n \rightarrow \, ^{32}_{15}P + ?$
26. $^{27}_{13}Al + ? \rightarrow \, ^{27}_{12}Mg + \, ^{1}_{1}H$

27. Check the incorrect statement:
   The valence of an element is equal to
   a. the number of electrons gained or lost by the element in forming a compound.
   b. the number of electrons shared by the reacting elements.
   c. the charge on the ion formed by the element.
   d. the difference between the number of protons and the number of neutrons in the nucleus.

    e. the difference between the number of protons in the nucleus and the number of electrons outside the nucleus.

28. A chemical bond formed by the sharing of electrons between the reacting atoms is known as
    a. An ionic bond
    b. A covalent bond
    c. A polar bond
    d. A dative bond
    e. An electrovalent bond

29. Balance the following equation:
$$\underline{\quad} C_2H_6 + \underline{\quad} O_2 \rightarrow \underline{\quad} CO_2 + \underline{\quad} H_2O$$
The sum of all the coefficients (including the coefficient of 1, when understood) in the balanced equation is
    a. 14
    b. 17
    c. 19
    d. 24
    e. Some other value

30. Balance the following equation:
$$\underline{\quad} AlCl_3 + \underline{\quad} AgNO_3 \rightarrow \underline{\quad} Al(NO_3)_3 + \underline{\quad} AgCl$$
The sum of all the coefficients is
    a. 6
    b. 8
    c. 10
    d. 12
    e. Some other value

31. Balance the following equation:
$$\underline{\quad} CaSO_4 + \underline{\quad} K_3PO_4 \rightarrow \underline{\quad} Ca_3(PO_4)_2 + \underline{\quad} K_2SO_4$$
The sum of all the coefficients is
    a. 9
    b. 8
    c. 6
    d. 4
    e. Some other value

32. How many grams of potassium chlorate must be decomposed to form 10 grams of oxygen?
    a. 18.3
    b. 19.7
    c. 23.5
    d. 25.5
    e. 51.4

33. The equation for the burning of methane, $CH_4$, is

$$CH_4 + 2O_2 \rightarrow CO_2 + 2H_2O$$

What volume of carbon dioxide, $CO_2$, will be produced by burning 8 grams of $CH_4$?

    a. 11.2 l.

    b. 22 l.

    c. 22.4 l.

    d. 44 l.

    e. Some other value

34. What percent of dry air is oxygen?

    a. 1%

    b. 21%

    c. 50%

    d. 75%

    e. 88%

35. Oxygen is NOT present in

    a. Water

    b. Gasoline

    c. Air

    d. Earth's crust

    e. Glass

36. Who discovered oxygen?

    a. Avogadro

    b. Boyle

    c. Cavendish

    d. Lavoisier

    e. Priestley

37. Oxygen is prepared in the laboratory by

    a. Decomposing water by heat

    b. Heating sand

    c. Heating potassium chlorate

    d. Heating nonmetallic oxides

    e. Reacting an acid with a base

38. An example of combustion is

    a. Corrosion of metals

    b. Decay of organic matter

    c. Rusting of iron

    d. Burning of a match

    e. Oxidation of food in the body

39. Which one is NOT a fire extingushing agent?

    a. Alcohol

    b. Carbon tetrachloride

    c. Liquid carbon dioxide

    d. Water

    e. Foamite

40. Check the incorrect statement:

a. Ozone is more active than oxygen.

b. Ozone is produced by the electrolysis of water.

c. Ozone has a pungent odor.

d. Ozone is used for deodorizing, disinfecting, and bleaching purposes.

e. Ozone is an allotropic form of oxygen.

41. In the reaction $Zn + H_2SO_4 \rightarrow ZnSO_4 + H_2$, which element is oxidized?

    a. Zn
    b. H
    c. S
    d. O
    e. None of these

42. In the reaction $2Na^+I^- + Cl_2 \rightarrow 2Na^+Cl^- + I_2$

    a. $Na^+$ is oxidized, $I^-$ is reduced.
    b. $Na^+$ is reduced, $I^-$ is oxidized.
    c. $I^-$ is oxidized, $Cl_2$ is reduced.
    d. $I^-$ is reduced, $Cl_2$ is oxidized.
    e. No oxidation-reduction takes place.

43. Underline the oxidizing agent in each of the following reactions in which oxidation occurs:

    a. $CO + O_2 \rightarrow 2CO_2$
    b. $NaOH + HCl \rightarrow NaCl + H_2O$
    c. $SnCl_2 + 2FeCl_3 \rightarrow SnCl_4 + 2FeCl_2$
    d. $Zn + Cl_2 \rightarrow ZnCl_2$
    e. $NaI + Br_2 \rightarrow NaBr + I_2$

44. Underline the reducing agent in each of the following reactions in which reduction takes place:

    a. $Fe_3O_4 + 4H_2 \rightarrow 3Fe + 4H_2O$
    b. $2HCl + Zn \rightarrow ZnCl_2 + H_2$
    c. $H_2SO_4 + Mg(OH)_2 \rightarrow MgSO_4 + 2H_2O$
    d. $SO_2 + 2H_2S \rightarrow 3S + 2H_2O$
    e. $2FeCl_3 + SnCl_2 \rightarrow 2FeCl_2 + SnCl_4$

45. Check the incorrect statement:

    a. Water reacts with soluble metallic oxides to form bases.
    b. Water reacts with nonmetallic oxides to form acids.
    c. Water decomposes when heated to a high temperature.
    d. Water forms hydrates with certain anhydrous salts.
    e. Water reacts with active metals, liberating hydrogen.

46. Check the incorrect statement:

    a. All hard waters can be softened by heating.
    b. All hard waters can be softened by the addition of sodium carbonate.
    c. All hard waters can be softened by passing through zeolite.

d. All hard waters can be softened by soap.

e. All hard waters can be softened by passing through ion-exchange resins.

47. Which of the following will react with water to give an acidic solution?

    a. $H_2$

    b. $SO_3$

    c. CaO

    d. $NH_3$

    e. $Na_2CO_3$

48. Which ion will precipitate soap in solution?

    a. $Cl^-$

    b. $Ca^{++}$

    c. $Na^+$

    d. $HCO_3^-$

    e. $NH_4^+$

49. Which one is NOT a factor affecting the solubility of a solid in water?

    a. Nature of the solute

    b. Temperature

    c. Pressure

    d. Surface of the solid

    e. Stirring or agitation

50. Check the incorrect statement:

A nonvolatile solute dissolved in water affects water in the following ways:

    a. Lowers its vapor pressure

    b. Raises its boiling point

    c. Lowers its freezing point

    d. Lowers its osmotic pressure

    e. Raises its density

51. A normal solution is one that contains

    a. One gram-molecular weight of solute per liter of solution.

    b. One gram-molecular weight of solute per liter of solvent.

    c. One gram-equivalent weight per liter of solution.

    d. One gram-equivalent weight per liter of solvent.

    e. One gram-equivalent weight per 1,000 grams of solution.

52. The gram-molecular weight of $H_2SO_4$ is 98. A solution containing 4.9 grams per 100 ml. of solution is

    a. 1M

    b. 0.2M

    c. 1N

    d. 0.2N

    e. 0.5N

53. What is the normality of a sulfuric acid solution containing 73.5 grams of $H_2SO_4$ per 500 ml. of solution?
   a. 1.5N
   b. 2.5N
   c. 3.0N
   d. 5.98N
   e. 6.0N

54. How many grams of NaOH are needed to make 500 ml. of a 1.5M solution?
   a. 20
   b. 30
   c. 60
   d. 75
   e. 120

55. Which of the following when added to blood will cause the red cells to burst?
   a. Distilled water
   b. Concentrated brine
   c. Concentrated glucose solution
   d. Physiological saline
   e. Plasma expanders

56. Classify the following as strong electrolyte (S), weak electrolyte (W), or nonelectrolyte (N):
   _____ a. Sugar
   _____ b. Acetic acid
   _____ c. Sodium hydroxide
   _____ d. Alcohol
   _____ e. Hydrochloric acid

57. Ammonium hydroxide is a weak base because
   a. It does not ionize.
   b. It has a small percentage of ionization.
   c. It will not react with an acid.
   d. It forms hydrogen ions in solution.
   e. One mole of ammonium hydroxide will not neutralize one mole of an acid.

58. Sodium acetate will hydrolyze to give a solution that is
   a. Acidic
   b. Basic
   c. Neutral
   d. Saturated
   e. Below pH 7

59. Which of the following 0.1N solutions has a pH slightly below 7?
   a. HCl
   b. NaOH

    c. NaCl

    d. $NaHCO_3$

    e. $NH_4Cl$

60. What is the pH of a 0.0001N NaOH solution?

    a. $-4$

    b. 4

    c. $-10$

    d. 10

    e. $10^{-10}$

61. The pH of acid gastric juice would be

    a. 0

    b. Below 7

    c. 7

    d. Above 7

    e. 14

62. A solution made up of a weak acid and its salt or a weak base and its salt, which resists appreciable changes in pH, is called a

    a. Normal solution

    b. Isotonic solution

    c. Neutral solution

    d. Buffer solution

    e. Colloidal solution

63. Based on the activity series: K, Ca, Mg, Zn, H, Cu, Hg, Ag, which of the following pairs of substances may be expected to react?

    a. Calcium chloride and hydrogen

    b. Dilute sulfuric acid and copper

    c. Silver nitrate and zinc

    d. Magnesium sulfate and mercury

    e. Potassium chloride and copper

64. Which one of the following reactions will NOT take place?

    a. $Zn + 2HCl \rightarrow ZnCl_2 + H_2$

    b. $Fe + CuSO_4 \rightarrow FeSO_4 + Cu$

    c. $Cu + 2HCl \rightarrow CuCl_2 + H_2$

    d. $3K + AlCl_3 \rightarrow 3KCl + Al$

    e. $Ni + 2AgNO_3 \rightarrow Ni(NO_3)_2 + 2Ag$

65. Which of the following will react with dilute sulfuric acid to form hydrogen and a soluble sulfate?

    a. Au

    b. Cu

    c. Hg

    d. Ag

    e. Zn

66. Check the correct statement:

Given one liter of 1N ammonium hydroxide and 1N sodium hydroxide

   a. Both solutions have the same pH.

   b. Both solutions contain the same number of grams of hydroxide.

   c. Both solutions contain the same number of $OH^-$ ions.

   d. In titration, both solutions would require the same amount of acid.

   e. The resulting titrated solutions would have the same pH.

67. Which of the following is NOT a property of acids?

   a. Turning red litmus blue

   b. Having a sour taste

   c. Yielding hydrogen ions

   d. Neutralizing bases

   e. Being proton donors

68. Which of the following is the correct name for KClO?

   a. Potassium chloride

   b. Potassium hypochlorite

   c. Potassium chlorite

   d. Potassium chlorate

   e. Potassium perchlorate

69. Assuming complete ionization, the pH of a 0.0001N solution of $Mg(OH)_2$ is approximately

   a. 0.0001

   b. 1

   c. 4

   d. 7

   e. 10

70. For complete neutralization, 1 liter of 1M NaOH would require

   a. 1 liter of 1N $H_2SO_4$.

   b. 2 liters of 1N $H_2SO_4$ solution.

   c. 1 liter of 1M $H_2SO_4$ solution.

   d. 1 gram-molecular weight of $H_2SO_4$.

   e. Some other value.

71. What volume of 0.200N $Ca(OH)_2$ would be required to neutralize 100 ml. of a 0.100N solution of HCl?

   a. 25 ml.

   b. 50 ml.

   c. 100 ml.

   d. 200 ml.

   e. 400 ml.

72. If it takes 12 ml. of 1N HCl solution to neutralize 20 ml. of NaOH solution, what is the normality of the NaOH solution?

    a. 0.24N
    b. 0.6N
    c. 1.77N
    d. 6N
    e. 24N

73. Check the incorrect statement:
    a. There are more organic compounds known than inorganic compounds.
    b. Organic compounds cannot be made from inorganic compounds,
    c. Organic compounds have lower boiling and melting points than inorganic compounds.
    d. Organic compounds are less ionic than inorganic compounds.
    e. Organic compounds are more complex in structure than inorganic compounds.

74. The one element that is present in all organic compounds is
    a. Oxygen
    b. Hydrogen
    c. Carbon
    d. Nitrogen
    e. Sulfur

75. Compounds that have the same molecular formula but different structural formulas are known as
    a. Allotropic forms
    b. Isomers
    c. Isotopes
    d. Polymers
    e. Homologs

76. Which is the molecular formula for a butene or butylene?
    a. $C_4H_4$
    b. $C_4H_5$
    c. $C_4H_6$
    d. $C_4H_8$
    e. $C_4H_{10}$

77. Gasoline is mainly a mixture of hydrocarbons of the
    a. Alkane series
    b. Alkene series
    c. Alkyne series
    d. Cycloalkane series
    e. Benzene series

78. In acetylene the carbon atoms are joined by
    a. An ionic bond
    b. A single bond
    c. A double bond

    d. A triple bond

    e. An electrovalent bond

79. Which one of the following formulas represents a member of the alkane or paraffin series?

    a. $C_3H_6$

    b. $C_4H_6$

    c. $C_5H_{12}$

    d. $C_6H_6$

    e. $C_{10}H_{20}$

80. Which compound is NOT an isomer of the other four?

    a. *n*-Pentane

    b. 2-Methylbutane

    c. 2,2-Dimethylpropane

    d. 2,3-Dimethylbutane

    e. Neopentane

81. Which reagent(s) will convert ethylene to ethyl bromide?

    a. $Br_2$ and $H_2O$

    b. HBr

    c. HBr and NaOH

    d. $CH_3Br$

    e. $CH_3Br$ and Na

82. Which one of the following reactions will NOT take place ordinarily?

    a. $CH_4 + Cl_2 \rightarrow CH_3Cl + HCl$

    b. $CH_2{=}CH_2 + Br_2 \rightarrow CH_2Br{-}CH_2Br$

    c. $CaC_2 + 2H_2O \rightarrow C_2H_2 + Ca(OH)_2$

    d. $CH_2{=}CH_2 + H_2O \rightarrow CH_3{-}CH_2OH$

    e. $C_5H_{12} + 8O_2 \rightarrow 5CO_2 + 6H_2O$

83. The best name for the compound shown below is

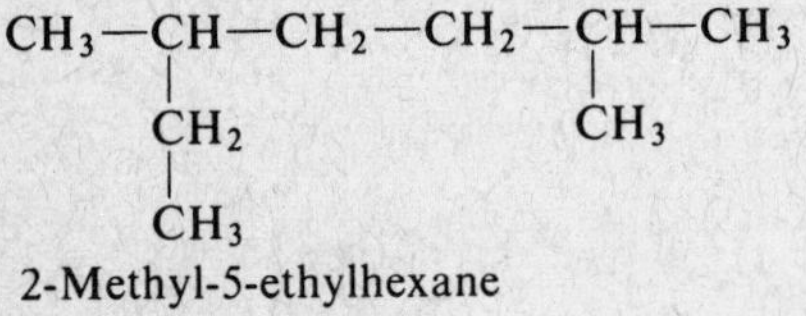

$$CH_3{-}CH{-}CH_2{-}CH_2{-}CH{-}CH_3$$
$$\quad\quad\ \ |\quad\quad\quad\quad\quad\quad\ |$$
$$\quad\quad CH_2\quad\quad\quad\quad\ CH_3$$
$$\quad\quad\ \ |$$
$$\quad\quad CH_3$$

    a. 2-Methyl-5-ethylhexane

    b. 2,5-Dimethylheptane

    c. 3,6-Dimethylheptane

    d. 1,1,4-Trimethylbutane

    e. 2-Ethyl-5-isopropylbutane

84. Check the incorrect pair:

    a. $CH_3Cl$, methyl chloride

    b. $CHCl_3$, chloroform

    c. $CH_2Br{-}CH_2Br$, ethyl bromide

    d. $CCl_2F_2$, Freon

    e. $CCl_4$, carbon tetrachloride

85. Which is an important use of Freon?
    a. Insecticide
    b. Drug
    c. Dry cleaning agent
    d. Refrigerant
    e. Plastic coating for frying pans

86. Check the incorrect statement:
    a. Methyl alcohol is produced by fermentation of sugars.
    b. Ethylene glycol is a common antifreeze for automobiles.
    c. Ethyl alcohol is present in all alcoholic beverages.
    d. Ethyl ether is used as a general anesthetic.
    e. Formaldehyde is used as a preservative for biological speci-
       mens.

87. Mild oxidation of a primary alcohol yields
    a. An ether
    b. An aldehyde
    c. A ketone
    d. A glycol
    e. Carbon dioxide and water

88. Secondary alcohols are oxidized to
    a. Aldehydes
    b. Ketones
    c. Acids
    d. Ethers
    e. Esters

89. Which is the correct name for $CH_3-\overset{\overset{\displaystyle O}{\|}}{C}-CH_3$?
    a. Ethyl acetate
    b. Methyl ethyl ether
    c. Secondary butyl alcohol
    d. Methyl ethyl ketone
    e. Acetone

90. Which are the products of reaction between $CH_3CH_2COOH$ and
    $CH_3OH$?
    a. $CH_3COOCH_2CH_3 + H_2O$
    b. $HCOOCH_2CH_2CH_3 + H_2O$
    c. $CH_3CH_2CH_2COOH + H_2O$
    d. $CH_3CH_2COOCH_3 + H_2O$
    e. $CH_3CH_2COOCH_2OH + H_2O$

91. Which compound undergoes hydrolysis to form acetic acid?
    a. Acetone
    b. Ethyl acetate
    c. Acetylene

    d. Methyl propionate

    e. Lactic acid

92. Check the incorrect statement:

    a. Acetic acid is present in sour milk.

    b. Formic acid is present in insect bites.

    c. Tartaric acid is present in grapes.

    d. Citric acid is a tricarboxylic acid.

    e. Esters are responsible for the flavors and fragrance of many fruits and flowers.

93. Alkaline hydrolysis of an ester is known as

    a. Esterification

    b. Saponification

    c. Addition

    d. Substitution

    e. Dissociation

94. A functional group common to both amines and amides is

    a. $-COOH$

    b. $-CHO$

    c. $-NH_2$

    d. $-NH_4$

    e. $-OH$

95. Which one of the following compounds is NOT an aromatic hydrocarbon?

    a. Benzene

    b. Toluene

    c. o-Xylene

    d. Ethylene

    e. Naphthalene

96. Which one of the following is NOT a common reaction of the aromatic hydrocarbons?

    a. Chlorination

    b. Bromination

    c. Nitration

    d. Sulfonation

    e. Addition

97. Which one of the following compounds is NOT a phenol?

    a. Carbolic acid

    b. Benzyl alcohol

    c. Picric acid

    d. p-Cresol

    e. Hexylresorcinol

98. Which one of the following is NOT related to aromatic amines?

    a. Aniline

    b. Acetanilide

    c. Sulfanilamide
    d. Phenacetin
    e. Trimethylamine

99. Which one of the following is NOT related to aromatic acids?
    a. Picric acid
    b. Aspirin
    c. Methyl salicylate
    d. Benzoic acid
    e. Salicylic acid

100. Which one is the preferred name for the compound shown below?

    a. *Ortho*-dichlorobenzene
    b. 2,5-dichlorobenzene
    c. *Meta*-dichlorobenzene
    d. Benzene dichloride
    e. *Para*-dichlorobenzene

101. Which of the following has never been used as a sulfa drug?
    a. Sulfanilic acid
    b. Sulfanilamide
    c. Sulfathiazole
    d. Sulfapyridine
    e. Sulfaguanidine

102. Which compound is an alkaloid?
    a. Niacin
    b. Quinine
    c. Adrenalin
    d. Aspirin
    e. Choline

103. Which of the following is an analgesic?
    a. Atropine
    b. Caffeine
    c. Morphine
    d. Reserpine
    e. Strychnine

104. Check the incorrect statement:
    a. Ribose is an aldopentose.
    b. Glucose is an aldohexose.
    c. Galactose is an aldohexose.
    d. Maltose is a ketohexose.
    e. Fructose is a ketohexose.

105. The principal sugar in the blood is
    a. Ribose
    b. Galactose
    c. Glucose
    d. Fructose
    e. Sucrose

106. The sugar that yields only glucose when hydrolyzed is
    a. Galactose
    b. Fructose
    c. Lactose
    d. Sucrose
    e. Maltose

107. Which sugar will NOT give a red precipitate of $Cu_2O$ when heated with Benedict's solution?
    a. Glucose
    b. Fructose
    c. Galactose
    d. Sucrose
    e. Maltose

108. Which salt of palmitic acid is a water-soluble soap?
    a. Barium palmitate
    b. Calcium palmitate
    c. Ferric palmitate
    d. Magnesium palmitate
    e. Sodium palmitate

109. The degree of unsaturation of a fat can be determined by means of
    a. Its iodine number
    b. Its saponification number
    c. The acrolein test
    d. The spot test
    e. Its melting point

110. Choline is a constituent of
    a. Lard
    b. Cottonseed oil
    c. Cholesterol
    d. Lecithin
    e. Methionine

111. Which element is NOT present in all proteins?
    a. C
    b. H
    c. O
    d. N
    e. P

112. Which of the following is NOT a test for proteins?

    a. Acrolein
    b. Millon
    c. Xanthoproteic
    d. Hopkins-Cole
    e. Biuret

113. Which of the following is NOT an amino acid?
    a. Leucine
    b. Valine
    c. Choline
    d. Lysine
    e. Alanine

114. The isoelectric point of a protein is the point (in terms of pH) at which the protein is
    a. The least soluble
    b. The most ionized
    c. The most electropositive
    d. The most electronegative
    e. The least denatured

115. Which of the following compounds has the peptide linkage?

    a. $CH_3-\overset{\overset{\displaystyle O}{\|}}{C}-OCH_2-CH_3$

    b. $H_2N-CH_2-\overset{\overset{\displaystyle O}{\|}}{C}-\overset{\overset{\displaystyle H}{|}}{N}-CH_2-COOH$

    c. $H_2N-\overset{\overset{\displaystyle O}{\|}}{C}-OCH_2-CH_3$

    d. $CH_3-\overset{\overset{\displaystyle O}{\|}}{C}-NH_2$

    e. $H_2N-\overset{\overset{\displaystyle O}{\|}}{C}-NH_2$

116. The name of the dipeptide shown below is

$$CH_3-\underset{\underset{\displaystyle NH_2}{|}}{CH}-\overset{\overset{\displaystyle O}{\|}}{C}-\overset{\overset{\displaystyle H}{|}}{N}-CH_2-COOH$$

    a. Alanylalanine
    b. Glycylglycine
    c. Alanylglycine
    d. Glycylalanine
    e. None of these

117. Which is the common name of the compound having the formula shown below?

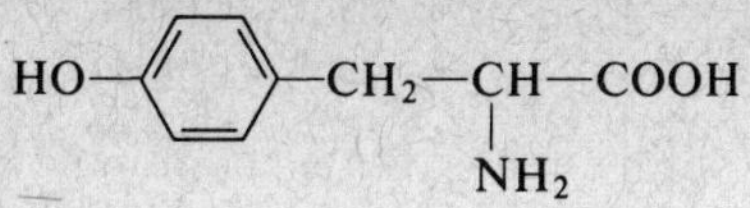

$$HO-\text{⬡}-CH_2-CH-COOH$$
$$| \quad NH_2$$

   a. Carboxylaminobenzene
   b. Phenylalanine
   c. Tryptophan
   d. Aminobenzoic acid
   e. Tyrosine

118. Which name is applied to the compound whose structure is shown below?

$$CH_3-S-CH_2-CH_2-CH-COOH$$
$$| \quad NH_2$$

   a. Tyrosine
   b. Methionine
   c. Tryptophan
   d. Phenylalanine
   e. Histidine

119. Which one is NOT a constituent of nucleic acids?
   a. Phosphoric acid
   b. Guanine
   c. Adenine
   d. Cytosine
   e. Alanine

120. Which of the following is a nucleoside?
   a. Cytosine
   b. Adenosine
   c. Adenosine monophosphate
   d. Adenosine diphosphate
   e. Adenosine triphosphate

121. Check the incorrect statement:
   a. Enzymes are specific in their actions.
   b. Enzymes are sensitive to heat.
   c. Enzymes are protein in nature.
   d. Enzymes are capable of initiating chemical reactions.
   e. Each enzyme has its specific optimum pH.

122. Which of the following is a coenzyme?
   a. Phosphatide
   b. Phosphocreatine
   c. Nicotinamide adenine dinucleotide
   d. Adenosine diphosphate
   e. Adenosine triphosphate

123. Which enzyme catalyzes the hydrolysis of proteins?
   a. Steapsin

    b. Amylopsin

    c. Insulin

    d. Ptyalin

    e. Trypsin

124. Which of the following converts trypsinogen to trypsin?

    a. Pepsin

    b. Chymotrypsin

    c. Enterokinase

    d. Secretin

    e. Gastrin

125. Which of the following reactions is catalyzed by amylopsin or pancreatic amylase?

    a. Starch to glucose

    b. Glycogen to glucose

    c. Sucrose to glucose and fructose

    d. Lactose to glucose and galactose

    e. Starch to maltose

126. Which of the following stimulates the evacuation of the gall-bladder?

    a. Enterokinase

    b. Cholecystokinin

    c. Secretin

    d. Gastrin

    e. Cholesterol

127. Which of the following is NOT produced by the pancreas?

    a. Amylopsin

    b. Steapsin

    c. Pepsinogen

    d. Trypsinogen

    e. Insulin

128. The activity of which gland affects the basal metabolic rate the most?

    a. Pancreas

    b. Thyroid

    c. Parathyroid

    d. Pituitary

    e. Adrenal

129. In which form is glucose stored in the liver?

    a. Lactic acid

    b. Lactose

    c. Ribose

    d. Glycogen

    e. Galactose

130. Which of the following accumulates in the muscle as a result of vigorous exercise?

a. Muscle glycogen
b. Lactic acid
c. Glucose
d. Carbon dioxide
e. Amino acids

131. Which of the following is a member of the ketone bodies?
    a. $\alpha$-Hydroxyacetic acid
    b. $\beta$-Hydroxybutyric acid
    c. Methyl ethyl ketone
    d. Acetophenone
    e. Benzophenone

132. Which of the following is NOT structually related to cholesterol?
    a. Vitamin A
    b. Vitamin D
    c. Sex hormones
    d. Bile salts
    e. Adrenal cortical hormones

133. Which of the following is called transamination?
    a. Conversion of amino acids to hydroxy acids
    b. Conversion of amino acids to keto acids
    c. Loss of ammonia from amino acids
    d. Formation of ammonium salts from ammonia
    e. Formation of urea from ammonia and carbon dioxide

134. Which amino acid is essential in the diet?
    a. Glycine
    b. Alanine
    c. Lysine
    d. Proline
    e. Serine

135. Which of the following compounds is NOT involved in urea synthesis?
    a. Ammonia
    b. Carbon dioxide
    c. Citrulline
    d. Ornithine
    e. Uric acid

136. Which is NOT a function of inorganic salts?
    a. Build and repair tissue.
    b. Influence the contraction of muscles and irritability of nerves.
    c. Maintain proper osmotic pressure.
    d. Maintain acid-base balance.
    e. Supply energy.

137. Which of the following inorganic salts is present in the body in only trace amounts?
    a. Iron

   b. Zinc
   c. Magnesium
   d. Sodium
   e. Potassium
138. Which of the following is NOT a factor in the clotting of blood?
   a. Thromboplastin
   b. Calcium ion
   c. Prothrombin
   d. Creatinine
   e. Fibrinogen
139. Which is the principal way in which carbon dioxide is carried in the blood?
   a. As carbonic acid
   b. As the bicarbonate ion
   c. As the carbonate ion
   d. As a carbamino compound with hemoglobin
   e. Adsorbed on the surface of the red cells.
140. Which compound is NOT a constituent of normal urine?
   a. Sodium chloride
   b. Albumin
   c. Urea
   d. Creatinine
   e. Uric acid
141. Which constituent of the urine remains relatively constant regardless of drastic changes in the diet?
   a. Ammonium salts
   b. Urea
   c. Uric acid
   d. Creatinine
   e. Inorganic phosphates
142. Which compound cannot be synthesized by the human body?
   a. Lactic acid
   b. Ascorbic acid
   c. Pyruvic acid
   d. Oleic acid
   e. Stearic acid
143. Deficiency of which vitamin is responsible for the disease beriberi?
   a. Riboflavin
   b. Pantothenic acid
   c. Niacin
   d. Thiamine
   e. Tocopherol
144. Which is NOT a member of the vitamin B complex?
   a. *p*-Aminobenzoic acid

    b. Ascorbic acid

    c. Folic acid

    d. Nicotinic acid

    e. Pantothenic acid

145. Which vitamin helps to prevent pellagra?

    a. Riboflavin

    b. Thiamine

    c. Nicotinic acid

    d. Pantothenic acid

    e. Folic acid

146. Which vitamin is formed in the body by exposure to ultraviolet irradiation or sunlight?

    a. Vitamin A

    b. Vitamin B

    c. Vitamin C

    d. Vitamin D

    e. Vitamin E

147. Which gland secretes the growth hormone?

    a. Pancreas

    b. Adrenal cortex

    c. Adrenal medulla

    d. Anterior pituitary

    e. Posterior pituitary

148. Which gland secretes the hormone that regulates calcium ion concentration?

    a. Parathyroid

    b. Thyroid

    c. Pituitary

    d. Adrenal

    e. Pancreas

149. Which hormone is NOT a steroid?

    a. Testosterone

    b. Insulin

    c. Cortisone

    d. Estradiol

    e. Progesterone

150. Which of the following is not related to the function of the thyroid gland?

    a. Toxic goiter

    b. Cretinism

    c. Dwarfism

    d. Myxedema

    e. Exophthalmic goiter

# ANSWERS TO TEST

| | | | | | |
|---|---|---|---|---|---|
| 1. | C, M, E, C, M | 37. | c | 67. | a |
| 2. | P, P, C, C, P | 38. | d | 68, | b |
| 3. | c | 39. | a | 69. | e |
| 4. | d | 40. | b | 70. | a |
| 5. | c | 41. | a | 71. | b |
| 6. | c | 42. | c | 72. | b |
| 7. | b | 43. | a. $O_2$; | 73. | b |
| 8. | c | | c. $2FeCl_3$; | 74. | c |
| 9. | d | | d. $Cl_2$; | 75. | b |
| 10. | S | | e. $Br_2$ | 76. | d |
| 11. | Br | 44. | a. $4H_2$; | 77. | a |
| 12. | Na | | b. Zn; | 78. | d |
| 13. | Ne | | d. $2H_2S$; | 79. | c |
| 14. | Ca | | e. $SnCl_2$ | 80. | d |
| 15. | b | 45. | c | 81. | b |
| 16. | D, A, E, B, C | 46. | a | 82. | d |
| 17. | c | 47. | b | 83. | b |
| 18. | e | 48. | b | 84. | c |
| 19. | d | 49. | c | 85. | d |
| 20. | b | 50. | d | 86. | a |
| 21. | d | 51. | c | 87. | b |
| 22. | $_{-1}^{0}e$ | 52. | c | 88. | b |
| 23. | $_{92}^{238}U$ | 53. | c | 89. | e |
| 24. | $_2^4He$ | 54. | b | 90. | d |
| 25. | $_1^1H$ | 55. | a | 91. | b |
| 26. | $_0^1n$ | 56. | N, W, S, N, S | 92. | a |
| 27. | d | 57. | b | 93. | b |
| 28. | b | 58. | b | 94. | c |
| 29. | c | 59. | e | 95. | d |
| 30. | b | 60. | d | 96. | e |
| 31. | a | 61. | b | 97. | b |
| 32. | d | 62. | b | 98. | e |
| 33. | a | 63. | c | 99. | a |
| 34. | b | 64. | c | 100. | e |
| 35. | b | 65. | e | 101. | a |
| 36. | e | 66. | d | 102. | b |

| | | |
|---|---|---|
| 103. c | 119. e | 135. e |
| 104. d | 120. b | 136. e |
| 105. c | 121. d | 137. b |
| 106. e | 122. c | 138. d |
| 107. d | 123. e | 139. b |
| 108. e | 124. c | 140. b |
| 109. a | 125. e | 141. d |
| 110. d | 126. b | 142. b |
| 111. e | 127. c | 143. d |
| 112. a | 128. b | 144. b |
| 113. c | 129. d | 145. c |
| 114. a | 130. b | 146. d |
| 115. b | 131. b | 147. d |
| 116. c | 132. a | 148. a |
| 117. e | 133. b | 149. b |
| 118. b | 134. c | 150. c |

# INDEX

# INDEX